EXILE, SCIENCE, AND *BILDUNG*

Studies in European Culture and History

edited by

Eric D. Weitz and Jack Zipes
University of Minnesota

Since the fall of the Berlin Wall and the collapse of communism, the very meaning of Europe has been opened up and is in the process of being redefined. European states and societies are wrestling with the expansion of NATO and the European Union and with new streams of immigration, while a renewed and reinvigorated cultural engagement has emerged between East and West. But the fast-paced transformations of the last fifteen years also have deeper historical roots. The reconfiguring of contemporary Europe is entwined with the cataclysmic events of the twentieth century, two world wars and the Holocaust, and with the processes of modernity that, since the eighteenth century, have shaped Europe and its engagement with the rest of the world.

Studies in European Culture and History is dedicated to publishing books that explore major issues in Europe's past and present from a wide variety of disciplinary perspectives. The works in the series are interdisciplinary; they focus on culture and society and deal with significant developments in Western and Eastern Europe from the eighteenth century to the present within a social historical context. With its broad span of topics, geography, and chronology, the series aims to publish the most interesting and innovative work on modern Europe.

Published by
Palgrave Macmillan: *Fascism and Neofascism: Critical Writings on the Radical Right in Europe*
by Eric Weitz

Fictive Theories: Towards a Deconstructive and Utopian Political Imagination
by Susan McManus

German-Jewish Literature in the Wake of the Holocaust: Grete Weil, Ruth Klüger, and the Politics of Address
by Pascale Bos

Turkish Turn in Contemporary German Literature: Toward a New Critical Grammar of Migration
by Leslie Adelson

Terror and the Sublime in Art and Critical Theory: From Auschwitz to Hiroshima to September 11
by Gene Ray

Transformations of the New Germany
edited by Ruth Starkman

Exile, Science, and *Bildung*

The Contested Legacies of German Emigre Intellectuals

EDITED BY
DAVID KETTLER
AND
GERHARD LAUER

EXILE, SCIENCE, AND *BILDUNG*

First published in 2005 by
PALGRAVE MACMILLAN™
175 Fifth Avenue, New York, N.Y. 10010 and
Houndmills, Basingstoke, Hampshire, England RG21 6XS
Companies and representatives throughout the world.

PALGRAVE MACMILLAN is the global academic imprint of the Palgrave Macmillan division of St. Martin's Press, LLC and of Palgrave Macmillan Ltd. Macmillan® is a registered trademark in the United States, United Kingdom and other countries. Palgrave is a registered trademark in the European Union and other countries.

ISBN 978-1-4039-6843-2

Library of Congress Cataloging-in-Publication Data

Exile, science, and Bildung : the contested legacies of German emigre intellectuals / edited by David Kettler and Gerhard Lauer.
   p. cm.—(Studies in European culture and history)
   Includes bibliographical references and index.
   ISBN 978-1-4039-6843-2 (alk. paper)
     1. Germans—United States—History—20th century. 2. Exiles—Germany. 3. United States—Intellectual life—20th century. 4. Germany—Emigration and immigration—History—20th century. 5. Political refugees—United States. 6. Refugees, Jewish—United States. 7. Brain drain—Germany—History—20th century. 8. Intellectuals—Germany—History—20th century. I. Kettler, David. II. Lauer, Gerhard. III. Series.

E184.G3E89 2005
303.48'273043'0904—dc22   2005047204

A catalogue record for this book is available from the British Library.

Design by Newgen Imaging Systems (P) Ltd., Chennai, India.

First edition: October 2005

10 9 8 7 6 5 4 3 2 1

CONTENTS

LIST OF ILLUSTRATIONS

LIST OF CONTRIBUTORS

JACK JACOBS. Professor of Government, John Jay College and The Graduate Center, City University of New York. Author: *On Socialists and the "Jewish Question" after Marx* (1992). Editor: *Jewish Politics in Eastern Europe: The Bund at 100* (2001).

LAURENT JEANPIERRE. Sociology. SHADYC (École des hautes études en sciences sociales, Marseille) / CEDITEC (Université Paris XII Val de Marne). Author: Articles on New York intellectuals, French exiles, Varian Fry, Surrealism, and History of the Social Sciences. Forthcoming book on the relationships between Surrealism and Structuralism.

DAVID KETTLER. Political Studies, Bard (Emeritus Trent University). Coauthor: *Karl Mannheim and the Crisis of Liberalism* (1995), *Social Regimes, Rule of Law and Democratic Change* (2001), and *Karl Mannheim's Sociology as Political Education* (2001).

GERHARD LAUER. Literature, University of Goettingen. Author: *Die Verspätete Revolution* (1995) and coeditor of volumes on problems of social history and literature, the return of the author.

REINHARD MEHRING. Institute of Philosophy, Humbold University, Berlin. Author: *Thomas Mann: Artist and Philosopher* (2001); *Introduction to Carl Schmitt* (2001).

GREGORY MOYNAHAN. History. Bard College. Dissertation: "The Face of the Times: Ernst Cassirer, Georg Simmel, and the Development of the Modern German Idea of Culture." Articles on Cassirer and Simmel.

ERNST OSTERKAMP. Literature. Humboldt University, Berlin. Author of numerous books and articles on German literature, with special interest in boundary between art and literature and the recent history of German "Bildung."

KAY SCHILLER. Modern Europan History. University of Durham, U.K. Author: *Gelehrte Gegenwelten. Über humanistische Leitbilder im 20. Jahrhundert* (2000) and articles on German–Jewish emigre historians in the United States as well as German history in the 1960s and 1970s.

ALFONS SÖLLNER. Political Theory. University of Chemnitz. Author, *Peter Weiss und die Deutschen. Die Entstehung einer politischen Ästhetik wider die Verdrängung* (1988); *Deutsche Politkwissenschaftler in der Emigration. Ihre Akkulturation und Wirkungsgeschichte* (1996); as well as some fifty scholarly articles on the social science emigration/return.

ANNA WESSELY. Art History and Sociology. Professor of Sociology, Eötvös Loránd University and Senior Fellow, Humanities Center, Central European University, Budapest. Editor of the *Budapest Review of Books*, author of articles on Simmel, Mannheim, eighteenth century philosophy, and contemporary art.

THOMAS WHEATLAND. Editorial Acquisition Department, Harvard University Press (Ph.D. in European and U.S. Intellectual History from Boston College, 2002). Author: Articles and a manuscript under development on the exile, history, and reception of the "Frankfurt School" in the United States.

IRVING WOHLFARTH. Professor of German Literature, Reims. Editor and contributor, *Nietzsche and an "Architecture of our Minds"* (1999); *Speak Of Camps; Think of Genocide* (1999).

JERRY ZASLOVE. Professor emeritus Literature and Humanities, Simon Fraser University, Vancouver, British Columbia Canada, monographs on Kafka, anarchism and culture, dialogic approaches to art and culture, recent publications on Siegfried Kracauer, Jeff Wall, Herbert Read and utopic modernism, and, forthcoming, on the death of the university, as well as an edited book of essays on Arnold Hauser.

PREFACE

The chapters in this volume originated in an ongoing project on intellectual exile at Bard College, directed by David Kettler, Research Professor.[1] They grew out of papers presented at a conference held at that college on August 13 to 15, 2002: "Contested Legacies: The German-Speaking Intellectual and Cultural Emigration to the United States and the United Kingdom, 1933–45." This conference in turn was prepared at the "No Happy End" workshop on February 13 to 15, 2001. The editors are accordingly indebted not only to the institution and donors whose contributions made these meetings possible, but also to the many colleagues who participated in them. Since there were twenty presenters at the workshop and fifty at the conference, it was obviously impossible to include all the high quality contributions in the present volume. Yet the project was seen from the outset to be a cumulative and collaborative effort. Accordingly, we would like to thank all the paper givers at both sessions who are not otherwise represented here: Peter Baehr,*** Reinhard Blomert, Jonathan Bordo,* Peter Breiner,*** Catherine Epstein, Christian Fleck,* Lawrence J. Friedman,* Judith Gerson, Lydia Goehr,* John Gunnell,** Wolfgang Heuer, Daniel Herwitz,* Claudia Honegger,* Martin Jay, Mario Kessler, Claus-Dieter Krohn,* Richard Leppert, Peter Ludes, John McCormick,* Neil McLaughlin, Berndt Nikolai, Margaret Olin, Hanna Papanek,* Paul Roazen, James Schmidt, Joanna Scott,* John Spalek, Michael P. Steinberg, Matthias Stoffregen, Edoardo Tortarolo, Roy Tsao, Mihaly Vajda, Suzanne Vromen,*** Wren Weschler, and Janet Wolff.[2]

Because of the intimate tie between the workshop, conference, and publications, we want to thank all the donors who supported the project in any of its phases: The German Academic Exchange Service (DAAD), The Max Kade Foundation, Inc., The Lucius N. Littauer Foundation, Inc., the Open Society Institute, James H. Ottaway Jr., and, at Bard, The Bard Center, the Human Rights Project, the Institute for International Liberal Education, and the Bard Music Festival. We are indebted for institutional support as well to the Seminar for German Philology of the Georg-August-University, Göttingen. Special thanks are owed to Leon Botstein, the President of Bard College, whose support extended from his guarantee of the workshop before there were any donors to his grant of a year's exemption from teaching, for work on this book, to one of the two editors, as well as the invitation to the Contested Legacy Conference to meet in the week framed by the weekends of the inspiring Bard Music Festival devoted to "Gustav Mahler and His World," of which he is the director.

Notes

1. The next phase of the project, "Limits of Exile," is introduced in David Kettler, " 'Et les émigrés sont les vaincus.' Spiritual Diaspora and Political Exile," *Journal of the Interdisciplinary Crossroads* (vol. 1, no. 2, August 2004).

2. Colleagues marked with an asterisk (*) are contributors to the collection of papers from the "No Happy End" Workshop, which also includes contributions by Laurent Jeanpierre, David Kettler, Ernst Osterkamp, Anna Wessely, and Jerry Zaslove, who are included in the present book: *Contested Legacies* (Berlin/Gleinicke: Galda & Wilch 2003). The double asterisk (**) marks the authors collected in a special issue of the *European Journal of Political Theory*, edited by David Kettler and Thomas Wheatland, which also contains an essay by Alfons Söllner: *Contested Legacies: Political Theory and the Hitler Era* (vol. 3, no. 2, April 2004). Independent publications of articles first presented at "Contested Legacies" include "Remigranten als Historiker in der frühen DDR," in Mario Kessler, *Exil und Nach Exil: Vertriebene Intellektuelle im 20. Jahrhundert*, Hamburg: VSA-Verlag, 2002, 181–197; David Kettler, " 'Weimar and Labor' as Legacy: Ernst Fraenkel, Otto Kahn-Freund, and Franz L. Neumann," Helga Schreckenberger, ed., *Die Alchemie des Exils. Exil als schoepferischer Impuls*, Vienna: Edition Praesens 2005; Margaret Olin, "The Road To Dura Europos," *Budapest Review of Books*, 12 (2002): 2–5; Janet Wolff, " 'Degenerate Art' in Britain: Refugees, Internees, and Visual Culture," *Visual Culture in Britain* (vol. 4, no. 2, 2003).

CHAPTER ONE

THE "OTHER GERMANY" AND THE QUESTION OF *BILDUNG*: WEIMAR TO BONN

David Kettler and Gerhard Lauer

The recognition of a difference between the scientific dimension of institutionalized knowledge in society and the rhetorical, didactic one, as well as the potential for conflict between them, is by no means unique to modern German culture. For centuries, English universities put the formation of clergymen and gentlemen ahead of the advancement of knowledge, and American colleges vied with each other in adapting both instruction and inquiry to the building of piety or moral character or civic virtue, not to speak of the utilitarian didactic achievements of inculcating commercial initiative or housewifely guile. Francis Bacon and Adam Smith denounced Oxford and Cambridge early in the modern era, and their spiritual heirs later created the London School of Economics, while the protests of Charles Beard and Thorstein Veblen against the higher education in America helped to bring into being the New School that was eventually to harbor an important contingent of the German émigrés of 1933.

Yet neither in England nor the United States did questions arising out of the contrasting aims of organized knowledge penetrate so deeply into competing designs of such knowledge, lay claim to such comprehensive ethical significance, resonate so profoundly in public discourses remote from debates about education in the narrower sense, or have such ambitions on the allocation of authority and power in society. Some of these themes doubtless arose among essayists elsewhere, as with Matthew Arnold or T. S. Eliot in Britain, or Ralph Waldo Emerson and Henry Thoreau in the United States, but the comprehensiveness, centrality, and pervasiveness of the problem constellation was distinctively German, as was its extension to spheres of discourse remote from the essayistic. The conception of Germany as uniquely a *Kulturnation* and of cultural policy consequently as the subject matter of prime political decisions was admittedly undermined by the defeat in World War I, which had been marked by this ideological motif, but in the world of the literary intelligentsia the conception revived in diverse forms during the Weimar years.

In the various discourses centered in the university faculties of philosophy, especially within the humanities and social studies, intellectual work in Germany was commonly scrutinized for its stand on the issues between *Bildung* (broad cultivation) and *Wissenschaft* (specialized research science), even if its substance was remote from pedagogical questions. The interrogation was a "philosophical" one, whether the writers sought to contribute to "orientation," to counter the loss of meaning widely associated with the explosion of modernity, or whether they were engaged in "specialist" science "for its own sake."

For the intellectuals forced into emigration by the Hitler regime, this dimension of their past intellectual activity, as well as the souvenirs of their participation in controversies about the supposed "crisis" of *Bildung* in the decades before 1933 remained a persistent presence. With their faces toward Germany, moreover, many of the émigrés grounded their claims to represent the "other," better Germany precisely on the charge that the Nazis had betrayed the *Bildung* ideal and practice that the emigration was safeguarding in exile. In their relations with English and American intellectual life, however, in the processes of acculturation that was a response to necessities as well as attractions, the older, "philosophical" context frequently appeared exaggerated and professionally unsound. These currents and countercurrents are differently managed by the émigré authors.

To add to the complexity of the situation, many of the elements of the core German *Bildung* tradition, the canonized names and poses, as well as the reputation of uncompromising German *Wissenschaft*, enjoyed high status in the significantly different setting of American campaigns against shallow moralism, commercialism, or hyper-specialization in higher education, notwithstanding the estrangement of the war years, 1917–1918. The high standing of German universities among American professors, especially of the older generation, was both evidenced and reinforced by the considerable number of them who had done a *Wanderjahr* of advanced study there, as a matter of course. Since late in the nineteenth century, moreover, the debate about American higher education was strongly influenced by conflicting citations of German models, a pattern of argument emphatically renewed by Abraham Flexner in his widely discussed *Universities—American, English, German*, published in 1930, on the eve of the post-1933 emigrations.[1] The exchange between Flexner and his critics offers a unique insight into the patterns of expectations—accepting or disparaging— that confronted émigré scholars, scientists, and intellectuals when they came, inescapably as Germans, to the American academic world.

The Academic Landscape in America: The Reception of Abraham Flexner's Idealization of German Universities

Flexner argued that neither American nor English institutions of higher education were more than secondary schools, in the last analysis, while Germany alone, building on the historic initiatives of Wilhelm von Humboldt, knew genuine universities. Above all, Flexner attacked the incorporation of vocational and "professional" training into the university. Only law may be included and medicine belongs, since these entail both rigorous scientific disciplines and humanitarian ideals. German students were brought to maturity, he contended, by their experience in the academic secondary schools, whose high standards were safeguarded by the nationwide *Matura* examination; and the universities were free to serve disciplined scholarship and science alone, without regard to the paternalistic or ad hoc utilitarian concerns of schools in the United States.

Flexner was an influential commentator at the time, an educationist whose power was by no means limited to the force of his public arguments. Although his retirement from his position as Secretary of the General Education Board—the Rockefellers' first educational philanthropy—was not altogether voluntary, he remained well connected with major donors in the field of education, respectful of his remarkable record. His proposals for massive reform in medical education, first in 1910 for the Carnegie Foundation for the Advancement of Teaching and later for the General Education Board, had been backed up by conditional foundation grants, whose terms he materially shaped, as was his scheme for a progressive secondary school, implemented in the Lincoln School at Teachers College, Columbia. To judge by the accounts in his autobiography, Flexner must have generated and programmed the expenditure of more than $60,000,000 on higher education during his years with Carnegie and Rockefeller. Within a year of the publication of his 1930 critique of American universities, moreover, he had been given the endowment funds to establish the Institute for Advanced Study in Princeton.[2] Even at the age of seventy, in short, he was a force that could not be ignored. He was in a unique position to renew public interest in the arguments based on idealized German school and university models that had been pushed aside by the ideological mobilization against Germany in World War I, the fear of Socialist influences from Germany in the postwar period, the distrust among social scientists and publicists of the "philosophical" and antiscientific motifs in German books such as Spengler's *Decline of the West*, and the celebration of new American models. After April 1933, then, he was also among the first to act on his admiration of the German academic tradition by assisting in the placement of distinguished émigrés.

Yet it would be a mistake, first, to confuse Flexner's thesis with an importation of the German debate about *Bildung* as it had developed during the Weimar years, with its presumed bearing on the philosophical aims and designs of knowledge, especially since he shows no awareness of the division between *Bildung* and *Wissenschaft* featured in the debate about the supposed "crisis of *Bildung*" in Germany. The actual Weimar debate compounded themes initiated in the nineteenth century by Nietzsche's assertions of the claims of "life" against the "dead knowledge" of the *Bildung* tradition, with ideas arising in the context of newly assertive social movements, to challenge the *Wissenschaften* at home in the universities, in a state of the question remote from the conjunction of the two concepts in the earlier idealistic ideology put forward in the name of Humboldt and still credited by Flexner. The debate about Max Weber's "Science as a Vocation" was the prime locus for this new turn in old arguments.

A number of the exiled intellectuals had been among those who sought for some mediation of the conflict, following the line laid down by Georg Simmel, whose authority outlived his death at the beginning of the epoch:

Anyone who has been active for decades in the academic sphere and who enjoys the trust of the youth knows how often it is precisely the inwardly most alive and idealistic young men who turn away in disappointment, after a few semesters, from what the university offers them in the way of general *Bildung*, the satisfaction of their innermost needs. For what they want, quite apart from the most outstanding instruction of a specialized and exact kind, is something more general or, if you like, something more personal. . . . Call this, if you like, a mere by-product of *Wissenschaft*, . . . but, if it is no longer offered to young people, the best among them will turn to other sources that promise to satisfy these deepest needs: to mysticism or to what they call "life," to social democracy or to literature in general, to a misunderstood Nietzsche or to a materialism tinged with scepticism. Let us not deceive ourselves. The German universities have largely surrendered the inner leadership of the youth to forces of this kind.[3]

Flexner knew nothing about this distinctive German theme of inner, subjective development, or about the conception of the "youth" as impatient, assertive actor in the struggle for *Bildung*. The emigrants brought these additional questions and expectations, as well as, for many of the Jews among them, the contradictory experience of the transmutation of *Bildung* from entryway into exclusionary formula, as *Bildung* had become a motto of anti-Enlightenment, *Gemeinschaft*-centered opinion in the course of the struggle with *Wissenschaft*.[4] Flexner's attempted revival of earlier American idealization of German university culture itself stands for a pattern of demands on the exiles that many of them will find puzzling and some will experience as demeaning, while others will use them as a route of access to academic standing.

Second, it would be an error to suppose that Flexner's undoubted capacity to gain attention for his theses about the superiority of the German universities meant that he could also redefine the field. The *Journal of Higher Education*, founded at the Ohio State University in the year that Flexner's book appeared, devoted its entire October, 1931 issue to reviews of Flexner's *Universities*; and these provide a valuable guide to American academic understandings of and responses to such challenges, as well as a preview of the context within which the German émigrés would have to find their way and their place a few years later.

The editor of the journal, W. W. Charters, may be excused the irritation expressed in his conclusion that Flexner only uses ridicule because he "has wholly missed the point" of university people trying to meet state-imposed obligations for professional training in vital social domains. Two of Charter's own studies are the object of almost three pages of such ridicule.[5] Charters grants Flexner that he has justly depicted the "Valhalla" of the research professor, but he denies that this addresses any of the real problems posed by the need to unite practice with knowledge, as a result of the "profound social forces which swept through the university to produce the professional schools." Another Ohio State professor, the philosopher, B. H. Bode, is more sympathetic to Flexner's critique of much American practice, but he finds a contradiction between, on the one hand, Flexner's abrupt disjuncture between secondary school *Bildung* and university research and, on the other, his insistence that university work too must be charged with "cultural values" and dedication to social intelligence. The failure to define either of those key concepts, as well as the contradiction itself, is, according to Bode, "apparently due to the fact that Mr. Flexner [. . .] takes over the [. . .] German conception of culture, lock, stock, and barrel." Once it is recognized that the unfinishable search for cultural values cannot be packaged in an authoritarian transmission of traditional ideas, as the Germans do, there is no further reason to draw Flexner's sharp line between school and university—and the liberal arts college comes back into its own. Bode closes with the ironic compliment that the book "may be expected to assist modern education in sloughing off a tradition from which the author himself has been unable to escape."[6]

Perhaps the most pointed criticism of Flexner's idealization of German models is made by the associate editor of the journal, W. H. Cowley, soon to be president of Hamilton College. In part, he is simply angry at Flexner's disdain of empirical research as a route to reform of higher education, but, more interestingly, he opposes Flexner as the main protagonist of "the German scholarly ideal," prevalent among graduate-studies-centered universities whose demands (and graduates) have ever more overshadowed "the traditional American ideal of the broad, symmetrical education of the individual." He speaks of mounting protest at the Association of American Colleges and increasing calls to action against the subordination of higher education to the purely intellectual interests of a tiny minority, at the sacrifice of the democratic requirement of an "enlightened citizenry."[7] A criticism similarly discomfited by what it takes to be Flexner's unreflective preference for a "feudal-aristocratic place of intelligence" rather than a democratic, practical one is especially noteworthy because it comes from William H. Kilpatrick, second only to Dewey in his importance for progressive education and a major figure in the development of the Lincoln School at Columbia Teachers College, which Flexner himself had originally brought into being. Unlike most of the other commentators, Kilpatrick takes note of Flexner's progressive strand, unparalleled in most German arguments, the suggestion that the autonomous research university is essential precisely because no other agency will provide the critical analysis of a "society that is driven it knows not whither by forces of unprecedented violence." Yet he objects that this is negated by Flexner's formalism, his "classical" penchant for neatly dividing school from university, *Bildung* from science, the learned from the rest. "Each idea and class must stand apart," Kilpatrick writes, "nicely bounded, not—as in democracy and modern logic—each one merging into its neighbor. Crude America must be withstood."[8]

Kilpatrick's criticisms gain added weight, from the standpoint of our present interest, when taken together with a review of Flexner by the leader of Kilpatrick's school, John Dewey, published in the spring of 1932. Sympathetic with Flexner's assaults against follies and distortions in higher education, he nevertheless protests that Flexner makes no attempt "to indicate the direction in which the American university might and should move." This can only be done, according to Dewey, if it is recognized that universities are a manifestation of the *ethos* of the national communities they serve.

> We—the American people—are blindly trying to do something new in the history of educational effort. We are trying to develop universal education; in the process we are forced by facts to identify a universal education with an education in which the vocational quality is pervasive. Mr. Flexner's criticisms would have been as truthful and as drastic if his criterion had been a recognition of what underlies both the excellencies and the defects of our society and our education instead of one which looks, however unconsciously, to the dualism of the past and of other societies.[9]

Dewey's judgment means that the American tendency that comes closest to the German insistence on the deep ethical and political ramifications of pedagogical arrangements is firmly committed to a uniquely American situation and mission in education.[10] Dewey and his associates, as democrats and humanitarians, will be among the leaders in welcoming the German émigrés, but they will also expect them to shift rapidly from the old to the new context of problems.

In sum, the intellectual and cultural émigrés from Germany entered an academic landscape where there were both avid friends and harsh critics of the specialist, research-centered, performance-oriented, autonomous German university system, but where neither the one nor the other actually grasped the state of the question of *Bildung*, as it was contested in the discourse of Weimar intellectuals, in the university and out.

Bildung as Contested Legacy

The present volume brings together a group of studies that variously explore the writings of several well-known members of the post-1933 German-speaking intellectual and cultural emigration against the background of the vicissitudes of *Bildung* in exile. Some twenty figures are included, ranging from Thomas Mann and László Moholy-Nagy to Erwin Panofsky and Paul Lazarsfeld, with special emphasis on the so-called neo-humanist tendency variously oriented to Ernst Cassirer or Thomas Mann, as well as the indispensable cultural commentators, including Adorno, Horkheimer, Benjamin, and Kracauer. That several individuals on this list were not themselves university faculty does not mean that any of them were removed from the *Bildung* controversy: the pervasiveness of the issue across the intellectual landscape is precisely the premise of the analysis. The aim is not, in any case, to rescue forgotten names but to explore an approach beyond the scope of past exile studies, to offer new help with the interpretation of texts and materials whose intrinsic value is not seriously in question. The idea is to read through the texts to the vicissitudes of *Bildung*, and then to reconsider the resulting transparent palimpsest. There are a number of insufficiently explored questions to be addressed by this means, ranging from the puzzling success of so many émigrés as university teachers to the no less puzzling sense of disappointment that haunted many of the émigrés who were most successful in making notable careers. At a more complex level of analysis, there are new insights to be gained, on the one hand, about the inner structure of the bargaining processes generally discussed as acculturation: the unfinished *Bildung* problem complex is usually neglected at the bargaining table. And, conversely, many obscurities or false notes in the writings of exile can be understood as documents of this practical aporia. Clearly, this issue is less likely to be present in cases where the condition of exile is effectively subsumed under patterns of international scientific migration, which were under way between Germany and the United States before Hitler came to power, supported in certain disciplines by the Rockefeller Foundation and other agencies dedicated to a global domain of *Wissenschaft*. These cases, especially in the natural sciences, have recently been made the subject of important studies, but they should not be overgeneralized.[11]

The individuals chosen for study here are members, with one or two exceptions, of what may be called the "Weimar generation," whose formative experiences came after World War I[12]; most are Jewish, at least by Nuremberg-law criteria, although at most one or two oriented themselves to any measurable extent to the internal Jewish debates about Jewish identity and culture; and almost all can be referred not only to one or another academic discipline but also to the more diverse congeries of cultural networks that was characterized in Germany as *die Intelligenz*. In the contests about *Bildung* that marked the Weimar period, none simply aligned themselves with a conservative defense of the nineteenth-century canon and ideal, according to the conventionalized forms of which they were all themselves schooled, but there were sharp differences among them as to the extent to which and the ways through which the ethical and political demands of the old *Bildung*, should be reconstructed. Issues in contestation included above all the relations between *Bildung* and Enlightenment, on the one hand, and between *Bildung* and (artistic) modernism, on the other. The former division entailed, in the German context, questions about democracy and the legitimacy of the Weimar state, and the latter, questions about antibourgeois revolution. Peter Nettl has characterized intellectuals as constituting "structures of dissent," in relation to the primary institutions of disciplinary knowledge, and this is adequate to our cases, as long as it is clearly understood that dissent is not limited to any given political direction, and that it may not take the form of political discourse at all.[13] In exile, arguably, nothing of this weighed as heavily as their common experiences and investments in a culture where the things they had learned, the things they studied, and the things they taught were widely believed to bear on the basic qualities of both individual and collective life. If nothing else, they had to explain a rejection of this culturist conception. Yet a number during the time of emigration hoped to influence events in Germany, perhaps to return, and several did so, in a few cases to positions of some prominence. The past and future of exile both figure in these studies of intellectuals in exile. The present-day literature on the concept and history of *Bildung* is large and accessible, and there is no need to retell here the story from Wilhelm von Humboldt to Eduard Spranger.[14] The aim is simply to map the controversy during the last Weimar years, with some attention to its concentrated form in pedagogical and curricular discussions.

Mapping the Weimar Dispute about *Bildung*

To complete our introduction, however, we first offer a schematic diagram of the intellectual landscape at the point of departure, a characterization of the major alignments in Weimar Germany on the question of relations between *Bildung* and *Wissenschaft*. The outer perimeter is defined by Max Weber, whose heroic abandonment of *Bildung* for the sake of *Wissenschaft*, most vividly in his address on science shortly before his death in 1920, is the limiting case by reference to which much of the Weimar debate proceeded.[15] Outside that boundary are scientific discourses whose self-reflection was confined to methodological issues and whose work served the *Bildung* debate largely as object lessons. Common to all of the authors enclosed by these bounds is a preoccupation with history, variously understood as both a prime constituent of and prime threat to *Bildung*. Yet the differences in treatment of that theme, generally subsumed under the heading of *Historismus*, do not lend themselves to easy classification, and they will be left for treatment in the separate studies.

Within the diagram, then, it is useful to plot four locations by matching two variables familiar from intellectual history, and then to attach a distinct plane defined by variable readings on a continuum of a different kind. On the four-fold table, then, one axis is labeled civilization and the other, politics. We separate thinkers whose concepts of *Bildung* are somehow reconciled with "civilization," a term often associated in contemporary discourse with the French Enlightenment and poised as antithesis against *Kultur*, from thinkers who are more true to the historical legacy of *Bildung* as an alternative to the supposed unhistorical "intellectualism" of civilization and the Enlightenment, and within each of these types, we distinguish, on the other axis, those whose conception of *Bildung* is expressly political from those who disdain conflictual politics. Representative of the upper left quadrant on our hypothetical table is the sociologist, Karl Mannheim; for the upper right, we take Hans Freyer, who is both sociologist and philosopher of culture; on the lower left, we locate authors like Ernst Cassirer and Ernst Robert Curtius; and the crowded lower right quadrant is represented by the various voices of the George Circle and "Secret Germany," as well as more pedantic voices of conservative opposition. It is perhaps emblematic of the discordant juxtapositions that are the subject of the present studies to force this complex assortment of intellectuals into the hostile confinements of so banal—and uncultivated—an analytical device. It is just as well, then, to disrupt the simplicity of the model with the addition of another dimension that is not susceptible to binary compartmentalization, the orientation to what may be called "revolutionary culturism," and that ranges from the unorthodox communist theorizing of the younger Georg Lukács, often revived by others during the Weimar years, with its reconceptualization of *Bildung* as class consciousness and its celebration of a revolutionary "new culture," at one end of the (dis)continuum, to various antinomian or anarchist articulations of avant-garde artistic rationales, no less scornful of Lukács's curious aesthetic conservatism than of the coordinated movements of his political associates. For present purposes, it will suffice to illustrate the four main alternatives on the principal dimensions, with emphasis on the quadrants that play a lesser part in the studies to follow but that remain an important part of the context.

Political Enlightenment as *Bildung*: Karl Mannheim

Our first striking marker, exemplifying a clear accommodation of both Enlightenment civilization and conflictual politics, is provided by a text arising in a specialist conference in 1932 devoted precisely to reinforcing the claims of sociology, a new and widely distrusted university discipline, to be accepted as a *Wissenschaft* in the desperate distribution struggles of the depression years. The speaker is one of the most polarizing—but also one of the most representative—figures of the age, Karl Mannheim. Probably drawing on the research of Hans Weil, whose book on the emergence of the concept of *Bildung* he selected for his own series, to appear immediately after *Ideologie und Utopie*,[16] and implicitly answering critics like Ernst Robert Curtius and Eduard Spranger, who charged him with betrayal of the German *Bildung* ideal, Mannheim offers a conciliatory account of the state of the question in 1932, just months before his own forced emigration:

> By specialized knowledge (*Spezialwissen*) we shall understand all the forms and contents of knowledge necessary for the solution of a scientific-technical or organizational task.

A knowledge whose advantage consists in its pure applicability and in its capacity of being separated from the purely personal is in essence always addressed to distinctively differentiated tasks in the social process, in a manner that is both particularistic and specialist. By *Bildung* knowledge (*Bildungswissen*), in contrast, we shall understand the tendency towards a coherent life-orientation, with a bearing upon the overall personality as well as upon the totality of the objective life-situation insofar as it can be surveyed at the time.[17]

Mannheim argued that sociology could not function as specialized *Wissenschaft* alone, and he maintained, audaciously—or, many thought, absurdly—that it was up to sociology in the present day to provide the "distinctive self-expansion of personality, together with the deepening of experiential dimensions, that was in large measure the meaning of the experience of *Bildung* for earlier generations."[18] If humanistic knowledge and the corresponding artistic culture were appropriate to the conditions of life of the defunctionalized aristocracy of the early nineteenth century and, in a different manner, of the passive and prosperous bourgeoisie later in the century, as Weil's study argued, then sociological self-understanding and practical orientation could meet the needs of democratized mass populations, especially through the *Bildung* of new meritocratic elites capable of stemming the slide to emotional mass democracy.

Mannheim is especially worth citing in this connection, first, because his example shows that the concern for *Bildung* was by no means limited to antimodern and politically conservative writers, as witness also the domestic political rationale for the *Hochschule für Politik*, which was the scene of operation for younger Jewish intellectuals, several among whom emerged as "political scientists" in emigration,[19] second, because he indicates that there was a common point to the concept, *Bildung*, despite conflict and fluidity about its contents during the Weimar years, and third, because he proposes, in effect, a peaceable division of labor between the two modes of knowledge, although he had no doubt as to which of the two complementary dimensions had the authority to draw the boundary lines. Common to all three elements is his determination to broker a deal between *Bildung* and Enlightenment, notwithstanding the enmity between them in a hundred years of *Bildung* discourse, an undertaking that was at one with his consistent support of the compromises constituting the Weimar constitution.[20]

Politics of the Will as *Bildung*: Hans Freyer

The second "political" compartment must be treated with subtlety, since it includes important thinkers who surprisingly paired the concepts of Enlightenment and *Bildung*, as Hegel had done in the *Phenomenology*, in order to decree that both are superceded. This is a motif, above all, in thinkers drawn to Martin Heidegger.[21] Yet this gesture cannot be taken as face value precisely because of the extent to which the thought is defined by the characteristic vision of a total disruption and renewal [*Aufbruch* und *Umbruch*] precisely in the domain that is more widely conceptualized as *Bildung*. The supposed rejection of *Bildung* was an opening to a reintroduction of its key elements.

In the political form of such "existentialism," Hans Freyer is a leading example. He deserves some careful attention in this introduction, precisely because the complex dialogic relationship between him and leading émigré thinkers is over-shadowed by the deep walls of separation erected by exile. Tactically allied with Mannheim against the proponents of a purely scientific sociology in 1932 but dia-metrically opposed to him on the bargain with the Enlightenment and Republic, Freyer pronounced his views on *Bildung* in a famous joint appearance with Carl Schmitt at a philosophers' congress in Davos in 1931. *Bildung* was obsolete, he pro-claimed, while promulgating a radically new regime of *Bildung*, based on a diagnosis of a failure to provide the communal conjunction (*Bindung*) that knowledge as rheto-ric should provide. Like Schmitt and Heidegger, who similarly attempted loyally—and failed—to inspire the Nazis with their related views of *Bildung*, Freyer was not divorced in Weimar from the intellectuals who subsequently became exiles. His aim is to outbid established "bourgeois" *Bildung* with a form that is revolutionary and all-encompassing, proclaimed under the name of "life."

Freyer denies that his conception of *Bildung* grounded on decision and will left *Wissenschaft* free, in the liberal manner, to pursue its autonomous way. Science is dis-tinguished only by its more rigorous method. Its grounding must be the same as *Bildung*. Ultimately, Freyer asserted, *Wissenschaft* had to educate students for practical activity. This meant a schooling of the will by spiritually deepening the force of decision. The old humanistic, bourgeois idea of *Bildung* (and the free-floating intelligentsia) had to be replaced by a political *Bildung* in which the person became rooted in the nation and was responsibly bound to the decision of the state. The old forms of education that focused on the totality of personality had to be replaced by those that disciplined the will for the tasks at hand. Students had to attain a sense of concrete duty, to be prepared to sacrifice, to dedicate their total person to nation and state.[22] The state in question, needless to say, could not be the pluralistic, Enlightenment-oriented constitutional regime of Weimar. Freyer is the second prime locational marker on our map.[23]

Bildung against "Politics": Ernst Robert Curtius and Eduard Spranger

Although the Weimar dispute about *Bildung* extended deep into the discourses of his-tory, philosophy, and philology, the challenges of sociology posed by both Mannheim and Freyer provided a central theme, above all because sociology figured so large in the cultural policies of the most influential universities minister of the Weimar era, Carl H. Becker.[24] Sociology, Becker thought, could provide the common civil under-standing that would enable individuals to recognize themselves through their deal-ings with others as peers and partners, without the discredited elitism and romanticism of the older conception. A sociological culture, moreover, would foster respect for diversity, as it encouraged individuals to take distance from themselves without fear of losing themselves. The individual subject of *Bildung* would reappear as a social being capable of being molded into a citizen; the universalistic assets of cul-ture would be recognized as elements of social cooperation; and the activism integral to *Bildung* would reveal itself as civic virtue. Leading roles in the public struggle against Becker's design were played by Georg von Below, a historian, Eduard Spranger, the philosopher best-known as an authority on Humboldt, and the noted Heidelberg critic of French literature, Ernst Robert Curtius.

The last-named is of special interest here, both because his devotion to French culture sharply separated him from the continuators of the anti-Enlightenment slogans of the "Spirit of 1914," and marked him as well, in the cultural politics of Weimar, as an unpolitical but comparatively moderate critic of the Republic, and because of the wider publicity of his views among non-university intellectuals.[25] Although the quadrant also includes neo-Kantian philosophers as well as classicists attentive to republican currents in ancient (and Renaissance) literatures, Curtius will serve as marker for the class, with the exposition supported by incidental comparisons with both Spranger and Mannheim.

Both Curtius and Spranger brought the discussion of the problematic relationship of *Bildung* to science and the sociopolitical sphere directly back to the source, since both first of all tried to assess the status of Humboldt's ideal in their time. Both endorsed Humboldt's emphasis on the need for organic harmony between individual freedom and supra-individual connections. Although Spranger, in accordance with historicist assumptions, wrote that the national culture was an individuality with a unified objective spirit, Curtius drew on the traditional humanism of the Rhineland instead of German historicism. Both believed the purpose of *Bildung* was the development of the individual as a cultural–ethical personality, who had to develop in the soil of objective value contents.[26] Curtius wrote: "We must return to the original foundation and beginning of our tradition and again learn the elements of culture."[27] With that specification as to the cultural source, he agreed with Spranger's contention that education was "the cultural activity that strives to bring about an unfolding of *subjective culture* in developing individuals, by means of an evaluatively guided contact with a given *objective culture* and the activation of a genuine, ethically requisite *cultural ideal.*" The larger "organic" totality relied on the *Bildung* of the individual for its realization. Spranger described this reciprocal relationship as the "infusion of the [objective] spirit with the [individual] soul and the infusion of the soul with the spirit. Where this succeeds in a productive sense, there is *Bildung.*"[28]

Spranger and Curtius sought to restore the nineteenth-century character of the university in the face of the new Weimar reform movement.[29] While they criticized the parliamentary democracy of the Republic to differing degrees, they were united in the belief that democratic forces had to be kept out of the university and that this meant putting the relatively new discipline of sociology in its proper place as a clear subordinate to traditional disciplines such as philosophy and history. Similarly, neither writer considered sociology as a kind of science that could be a fit partner for a coalition to restore *Bildung.* For Spranger, the rejection extended to all modern sciences. He wrote that *Wissenschaft* had come to mean a mere positivistic and utilitarian specialization oriented toward the adaptation of practical abilities to the material here and now. He saw sociology as the epitome of this orientation. In addition to its mechanistic methodology, it limited ethical questions to those of social forms. When this concern for the practical removed a will to values (*Wertwille*), he asserted, the result was relativism.[30]

Hostile to Becker's proteges and projects, Curtius and Spranger looked beyond the politicized Weimar state for a solution to the cultural crisis of Weimar. Spranger called for a renewal of culture carried by new forces, notably the youth movement, which he apostrophized in wholistic terms, without regard to the political character of the "renewal" movements he had in mind or of the deep cleavages within the youth cohort. The new culture would be based on a larger ideal tied to both the achievements of the past as well as to the meaning provided by religion.[31] Curtius in contrast shared Mannheim's conviction that the primary responsibility rested with the intellectuals, but he believed that their proper contribution was precisely the restoration of faith in the Western cultural heritage that Mannheim considered obsolete. The food for this *Bildung* was to be humanism reinspired by the Renaissance, a humanism that came from the creative intensity of life and was connected with religious belief. Warning that Germany should not make an abrupt break with the past, Curtius also looked to the academic tradition. The more democracy brought the masses to the fore, the greater the need for a restored humanistic elite and its secure field of operation in the university would become.[32]

In this connection, then, Curtius gave voice to his hostility to Jews who do not choose either the route of full assimilation to the host culture or wholehearted traditionalism. "We hope that youth—German youth—will resist all attempts by scientific authorities," Curtius continues, "to dissuade them from an appreciation of greatness and idealism."[33] It is not necessary to question the sincerity of Curtius's rejection of racialist anti-Semitism or his abhorrence for the National Socialists, to emphasize the intimate connections between even this moderate and comparatively modernist evocation of the older *Bildung* tradition with furiously anti-Socialist and anti-Jewish themes, a consideration doubtless of moment to many of the emigrants who had been otherwise sympathetic to Curtius's urbane literary approaches.

Above *Bildung* and Politics: The George Circle

Youth served as an even more militant slogan in the antipolitical and anti-Enlightenment *Bildungspolitik* of the fourth quadrant of our diagram, at least among the most representative figures, but it was a youth configured according to the leader-follower design pervasive in the prewar German youth movement and crystallized in the remarkably attractive George Circle and its extensive penumbra. Karl Mannheim wrote a compelling account of Heidelberg as a cultural center divided between the realms of Max Weber's sociology and the George Circle; and Erich Kahler, on the periphery of the circle, effectively launched his public career with an attack on Max Weber's *Wissenschaft als Beruf*, a publication that also played a key part in the disputes already surveyed.[34] Yet arguments about *Bildung* in the university hardly penetrated into the inner poetic circle.

The issues were translated to a distinctive sphere, ostensibly remote from the realms of politics or civilizational currents. The struggle was indeed about the *Bildung* of the youth, but the question of *Wissenschaft* was simply pushed aside. Although Max Weber may be seen to have asserted the heroism of an irresistible modernity devoid of "meaning" as the term figured in the "crisis" debate, a condition where science is a vocation without being a "calling" in some transcendent sense, George and his circle promised a calling to a secret and ultimate *Bildung*, outbidding all other invocations of the term. George's pronouncement, "From me, no road leads to *Wissenschaft*," cites the contrast central to the wider debate,[35] but he offers no more explanation of what he means by *Wissenschaft* than he does of *Bildung*. The only thing certain was that it would transform humanity. The inflated metaphors of "life," "youth," and the "organic" suggested to the chosen that they were already living in the predawn of that future. An "icon [*Bildnis*] of the master" seemed to suffice for their *Bildung*. *Bildung* required neither universities or other institutions nor any specified contents. *Bildung* was rather to resemble a religious epiphany, as was manifest in the discussions within the circle about correct iconic practices, as well as in the belief that the exclusivity of poetics could take the place of all *Wissenschaft* testify to this.[36] The *Bildung* of the few was modeled on the "circle" of Jesus' disciples, the reading of a poem was a neo-religious liturgy, and the abstinence from politics was a matter of principle.[37] It was precisely in this respect that they considered themselves superior to the "old" sciences, and their claims to a capacity for *Bildung*.

There were parallels to the George Circle, as with the poetic pretensions and the vast ambitions of Rudolf Borchard, which are not easy to distinguish from it at this distance in time. They all present themselves as unpolitical, but nevertheless seek to exercise a direct influence on the political through their elite "few." Of course, the political does not refer here to anything as mundane as social legislation or anti-inflationary policy. The political has *Bildung* as its aim, and it is supposed to draw in ever wider circles, beginning from "above" by means of the few. It is hard to avoid noticing the exaggerated self-importance of this project, but it is no less important to remark its extraordinary appeal to outstanding intellectuals at the time. Since it was impossible to speak of *Bildung* without simultaneously alluding to the educated middle class for whom the concept was iconic, it was necessary to widen the distance. They spoke of "unconditional renewal" and they were not economical with expressions of totality and the direct apprehension of essences. The specialization characteristic of modern orders of knowledge seemed to be suspended, and the "essential," to come into view.

With all this diffuse intensity, it is no wonder that the ways of the George followers could as easily terminate in the "new" state as in "inner emigration" or exile. The fault lines among the individual members could not be compellingly predicted, and it was not clear to the participants themselves. That the members of the circle returned to the Master's ideas after 1945 is further evidence of the striking persistence of the all-too German controversy about *Bildung*, from which this fourth quadrant on the intellectual landscape cannot be excluded.

An Overview of the Studies in this Volume

No group active during the last imperial years in Germany and the fourteen years of Weimar was more intimately identified with the slogan of *Bildung*, expressly treated as remote from specialized sciences of the university and as core of an intimate association of individuals, than the so-called circle around the poet, Stefan George. The members forced into exile after 1933 could not avoid decisions about this motif so demandingly present in their earlier lives, and they exemplify a significant range of alternative ways of managing this unshakable legacy, encapsulated in an elitist conception of "Secret Germany" that also resonated with many intellectuals drawn to National Socialism. In a brief essay grounded in his extensive scholarly investigations of George and his circle, *Ernst Osterkamp* evokes and analyses this extreme exile experience.

Irving Wohlfarth, in turn, closely marks Walter Benjamin's dialectical conversion of "Secret Germany" into a location defined by "political repression and the denial of a public voice." Although Benjamin's exile ended in death before he could join most of the individuals studied in this volume in one of the secure English-speaking nations, his years in Germany had been a constant preparation for exile, and the production of his brief years in French exile became a vital impulse in the self-confrontation of such disparate prominent exile figures as Hannah Arendt and Theodor Adorno. Yet, as Wohlfarth shows, Benjamin to an important extent kept his own secret, leaving it to be differently unriddled in different times. This *Bildung* has to be painfully extracted from the given materials; and it bestows no status. With all the hermeticism and rebelliousness of Benjamin's position, as Wohlfarth shows, he profoundly affects the debate about Enlightenment, politics, Jews—and Germany.

Reinhard Mehring presents Thomas Mann as a writer who brings these issues more into the light, challenging the thought that myth and humanism are contradictory, first of all through the humanistic prototypes that people his novels but secondly also in philosophical essays, whose arguments are explicated by several exile philosophers and in the course of Mann's correspondence with Theodor W. Adorno. With the end of exile, as such, Mann's humanistic *Bildungs* project was largely abandoned by those who had expounded it earlier, and his extraordinary authority abruptly withered, especially within the new German literary scholarship.

A former secretary and close friend of Thomas Mann stood out, according to *Gerhard Lauer*, as a lifelong witness to a distinctive idealization of a "revolutionary" refounding of German *Bildung*, having chosen Germany above his native Austria, which he deemed polluted by Habsburg. In principle, Kahler made no concessions to the intellectual currents of his place of asylum, except insofar as his self-popularizations served a taste for cultural uplift. Here was an exile in the vestibule.

The contrast could hardly be sharper than with another émigré who similarly came from former Hapsburg territory to Germany, the artist, Lázló Moholy-Nagy, who enjoyed the success of a cosmopolitan figure. *Anna Wessely* shows how he was moved to leave Germany, where he had earlier come as exile from Hungary and how the terms of his welcome in America nevertheless compelled him to abandon the *Bildung* project of the *Bauhaus*, the aim that was of special value to him.

Another group of avant-garde émigré artists illuminate a very different aspect of the American exile scene. According to *Laurent Jeanpierre*, the exiled surrealists are linked to the aesthetician of the Institute of Social Research by the question of acknowledging the inescapable inner connection between *Bildung* and myth without a romantic transcendence of practical enactment in the realm of concrete social possibilities, notwithstanding the failure of both to recognize the hidden parallels.

Gregory Moynahan moves the discussion to the inner philosophical structure of Ernst's Cassirer's ambiguous accomplishment in providing a context for negotiations between certain émigré thinkers and their American hosts. This reading requires a reconsideration of Cassirer's famous debate with Martin Heidegger in Davos in 1927. Cassirer's further elaboration of his argument—and its application to large political themes—shows clear and influential marks of his engaged encounter with the philosophical setting in his places of exile, and his service as mediator in the transition to the new frame of reference is especially clear in the work and reception of Erwin Panofsky.

An unexpected affinity can be observed, according to *Kay Schiller*, between Cassirer and the Renaissance scholar, Paul Oskar Kristeller, whose rigorous historical method precluded in principle the more symbolic rendering of humanist ideas that Cassirer occasionally ventured in his writings in exile. At issue in the "humanistic turn" was the transfer to the United States of the debate about a "Third Humanism" that had taken authoritarian and elitist form in Germany. Cassirer and Kristeller came together in insisting on the philosophical seriousness of the Renaissance figures they studied and in rejecting their invocation in aid of stereotyped positions in the old *Bildungs* debate.

In his extended and extensively researched essay on Siegfried Kracauer, *Jerry Zaslove* brings forward the profound soberness with which this essayist, a convenor and participant of the Weimar *Bildung* debate, explores the experience of exile—and Holocaust—notably in displacing the destructive and wordy yearning for community with the distanced but infinitely attentive view of the photographer. Interestingly, Kracauer's work was completed, not by Theodor Adorno, with whom he had been in extended conversation, but by Paul Oskar Kristeller, who evidently found him a kindred spirit precisely in this prophetic *ataraxia*.

Jack Jacobs contends that the stark recognition of anti-Semitism as focal point of attention by the Horkheimer-Adorno group, during the years of exile, mediated the displacement of the theoretical focal point from Marxist social theory to a distinctive dialectical cultural–political configuration. Not only their reflections on the German spectacle, seen from a distance, but also the exigencies of their client search and service contributed to this reorientation. The conditions of life in exile manifested itself on more than one level.

The complexity of the exile situation of this exemplary group is made evident by an account of their anti-Semitism research project from *Thomas Wheatland's* competing perspective, where the emphasis is rather on the effort to meet the methodological expectations of American social science clients. The emphasis here is on the contested importance of the empirical methodologist—and fellow-exile—Paul Lazarsfeld for their work, on issue of particular interest because it highlights the question of the extent to which participants in exile revised their memories in the changed contexts of later times.

The last chapter, like the earlier one on Thomas Mann, moves on to the end of exile. The conjunction of politics and culture, both in the diagnosis of the crisis and in the projection of its negation, was central to Adorno, *Alfons Söllner* shows, in the *Bildungs* practice in Germany, to which Adorno returned after the years of exile. Adorno's strategy of "political culturism" is a sign of the breach produced by exile and of the subtlety required to work effectively across the gap.

Notes

1. Abraham Flexner, *Universities—American. English. German* (New York, London, Toronto: Oxford University Press, 1930). For the composition and reception of the book, see Thomas Neville Bonner, *Iconoclast. Abraham Flexner and a Life of Learning* (Baltimore, London: The Johns Hopkins Press, 2002), 213–233 (hereafter cited as *Iconoclast*). For a historical overview of the "importation of [. . .] German ideals of university education," see Jürgen Herbst, "Liberal Education and the Graduate Schools: An Historical View of College Reform," *History of Education Quarterly* 2, 4 (December 1962): 244–258.

2. Abraham Flexner, *I Remember. The Autobiography of Abraham Flexner* (New York: Simon and Schuster, 1940) (hereafter cited as *I Remember*).
3. Klaus Lichtblau, *Kulturkrise und Soziologie um die Jahrhundertwende: Zur Genealogie der Kultursoziologie in Deutschland* (Frankfurt: Suhrkamp, 1996), 408. Cf. Karl Mannheim, "Science and Youth," in Karl Mannheim (ed.), *Sociology as Political Education* (New Brunswick: Transaction, 2001), 99–104 (Volume cited hereafter as *Sociology as Political Education*).
4. For an excellent analytical overview, see Aleida Assmann, *Arbeit am nationalen Gedächtnis. Eine kurze Geschichte der deutschen Bildungsidee* (Frankfurt, New York: Campus-Verlag, 1993), 72–77, 85–91 (hereafter cited as *Arbeit am nationalen Gedächtnis*).
5. W. W. C. [Werrett Wallace Charters] "Editorial Comments," *The Journal of Higher Education* 2, 7 (October 1931): 104.
6. B. H. Bode, "Currents and Cross-Currents in Higher Education," *The Journal of Higher Education* 2, 7 (October 1931): 374–379. David Snedden similarly chides Flexner for fixing on an obsolete pattern. Writing in an evolutionist mode, Snedden suggests that the changes that Flexner decries may well be experimental steps toward a realization of Flexner's high ideals in a manner appropriate to emerging social conditions. ("Functions of the University," *Journal of Higher Education* 2, 7 (October 1931): 384–389.)
7. W. H. Cowley, "The University and the Individual. The Student as an Individual in Contrast to Mr. Flexner's Interest in his Intellect Only," *The Journal of Higher Education* 2, 7 (October 1931): 390–396. A defense of the undergraduate college with a special emphasis is offered by Henry N. Maccracken of Vassar College, writing on "Flexner and the Woman's College," *The Journal of Higher Education* 2, 7 (October 1931): 367–373.
8. William H. Kilpatrick, "Universities: American, English, and German," *The Journal of Higher Education* 2, 7 (October 1931): 357–363. In the early 1930s, Kilpatrick and his associates were concerned above all with the conjunction between progressive education and radical social change, and the German connection tended to take the form of collaboration with German critics of the existing *Bildung* regime. See Herman Röhrs, "Progressive Education in the United States and its Influence on Related Educational Developments in Germany," in Jürgen Heideking, Marc Depaepe, Jürgen Herbst (eds.), *Mutual Influences on Education: Germany and the United States in the Twentieth Century. Paedagogica Historica* 33, 1 (1997): 45–68.
9. John Dewey, "[Book Review of] Universities: American, English, German. By Abraham Flexner," *International Journal of Ethics* 42, 3 (April 1932): 331–332. Bonner reports on a subsequent exchange of letters between Dewey and Flexner, which ends with expressions of mutual good will but no basic agreement on the question of relating cultural and vocational aims, not to speak of Dewey's insistence on a standard specific to the American social context. Bonner, *Iconoclast*, 232–233.
10. Cf. David A. Hollinger, *Science, Jews, and Secular Culture* (Princeton: Princeton University Press, 1996).
11. Mitchell G. Ash and Alfons Söllner (eds.), *Forced Migration and Scientific Change. Emigré German-Speaking Scientists and Scholars after 1933* (Washington: German Historical Institute, Cambridge University Press, 1996); Giuliana Gemelli (ed.), *The "Unacceptables." American Foundations and Refugee Scholars between the Two Wars and after* (Brussels et al.: Peter Lang, 2000).
12. Karl Mannheim, "The Problem of Generations," in Karl Mannheim (ed.), *Essays in the Sociology of Knowledge*, ed. and trans. Paul Kecskemeti (London: Routledge and Kegan Paul, 1952): 276–322. Cf. David Kettler and Colin Loader, "Temporizing with Time Wars: Karl Mannheim and Problems of Historical Time," *Time and Society* 13, 2 (September 2004).
13. Peter Nettl, "Ideas, Intellectuals, and Structures of Dissent," in Philip Reif (ed.), *On Intellectuals* (New York: Doubleday & Company, 1969), 57–134.

14. Aleida Assmann, *Arbeit am Nationalen Gedächtnis*; Georg Bollenbeck, *Bildung und Kultur. Glanz und Elend eines deutschen Deutungsmusters* (Frankfurt, Leipzig: Insel, 1994); Walter H. Bruford, *The German Tradition of Self-Bildung: "Bildung" from Humboldt to Thomas Mann*. (Cambridge: Cambridge University Press, 1975); Leonard Krieger, *The German Idea of Freedom* (Chicago: University of Chicago Press, 1957); Fritz K. Ringer, *The Decline of the German Mandarins; The German Academic Community, 1890–1933* (Cambridge/MA: Harvard University Press, 1969); Gunter Scholtz, *Zwischen Wissenschaftsanspruch und Orientierungsbedürfnis: Zur Grundlage und Wandel der Geisteswissenschaften* (Frankfurt: Suhrkamp, 1991); Heinz-Elmar Tenorth, *Zur deutschen Bildungsgeschichte* (Köln, Wien: Böhlau, 1985); Hans Weil, *Die Entstehung des deutschen Bildungsbegriffs* (Bonn: Cohen, 1930 [Bonn: Bouvier, 1967]) (hereafter cited as *Die Entstehung des deutschen Bildungsbegriffs*).
15. A companion to the present volume is a special edition of *The European Journal of Political Theory* 3, 2 (2004). "Contested Legacies: Political Theory and the Hitler Era," David Kettler and Thomas Wheatland (eds.), devoted to conflicts over Max Weber's legacy in exile, as well as to several studies of Hanna Arendt. Both publications are based on papers presented to a conference held at Bard College in August 2002, "Contested Legacies: The German-Speaking Intellectual and Cultural Emigration to the US and UK, 1933–1945."
16. Weil, *Die Entstehung des deutschen Bildungsbegriffs*.
17. Karl Mannheim, "The Contemporary Tasks of Sociology: *Bildung* and the Curriculum," in David Kettler and Colin Loader (eds.), *Sociology as Political Education* (New Brunswick, London: Transaction, 2001). A translated extract from Karl Mannheim, *Gegenwartsaufgaben der Soziologie* (Tübingen: J.C.B. Mohr, 1932) (hereafter cited as "The Contemporary Tasks of Sociology"). For background and commentary, see Colin Loader and David Kettler, *Karl Mannheim's Sociology as Political Education* (New Brunswick and London: Transaction, 2002) (hereafter cited as Karl Mannheim's Sociology as Political Education).
18. Mannheim, "The Contemporary Tasks of Sociology," 155.
19. Tutors included Hans Morgenthau, Franz L. Neumann, and Ernst Fraenkel, but the most important voice was Albert Salomon. See Albert Salomon, "Innenpolitische *Bildung*," in Ernst Jäckh (ed.), *Politik als Wissenschaft. Zehn Jahre Deutsche Hochschule für Politik* (Berlin: Hermann Reckendorf, 1934), 94–110; see Ulf Matthiesen, " 'Im Schatten einer endlosen großen Zeit.' Etappen der intellektuellen Biographie Albert Salomons," in Ilja Srubar (ed.), *Exil. Wissenschaft. Identität. Die Emigration deutscher Wissenschaftler, 1933–1945* (Frankfurt: Suhrkamp, 1988), 299–350.
20. Mannheim chose the language of "synthesis" rather than negotiations, but his actual exposition of synthesis justifies the analytical terminology chosen, especially because of its congruence with precisely the most despised feature of the Weimar project, its desperately unlucky but thoroughly admirable option for pluralism. See David Kettler and Volker Meja, *Karl Mannheim and the Crisis of Liberalism* (New Brunswick and London: Transactions, 1995). For a less political reading of this connection and a definitive analysis of Mannheim in the context of the philosophical problem constellation of his time, see now Reinhard Laube, *Karl Mannheim und die Krise des Historismus* (Göttingen: Vandenhoeck & Ruprecht, 2004) especially the final chapter, "*Bildung* als Perspektivierung," 500–540.
21. The theme is well developed in Karl Löwith, *Mein Leben in Deutschland vor und nach 1933* (Stuttgart: Metzler, 1986), 27–42 (hereafter cited as *Mein Leben in Deutschland vor und nach 1933*).
22. Hans Freyer, *Das politische Semester: Ein Vorschlag zur Universitätsreform* (Jena: Eugen Diederich, 1933), 8, 16–18, 37, 39.
23. With special thanks to Colin Loader, the exposition of Freyer's concept of *Bildung* draws heavily on Loader and Kettler, *Karl Mannheim's Sociology as Political Education*.

24. See Loader and Kettler, *Karl Mannheim's Sociology as Political Education*. The account of Eduard Spranger and Ernst Robert Curtius also draws extensively on this book.

25. Ernst Robert Curtius occasioned Karl Mannheim's only published defense of *Ideologie und Utopie* against his critics, and it focused exactly on his charge that "sociologism" betrayed the German ideal of *Bildung*. Notably, the exchange took place in *Neue Schweizer Rundschau*, a periodical aimed at a broad public of cultural intellectuals. Mannheim laid claim to the legacy of the recently dead great German teachers, Max Weber, Ernst Troeltsch, and Max Scheler. Ernst Robert Curtius, "Soziologie—und ihre Grenzen," *Neue Schweizer Rundschau* 22 (1929): 727–736, and Karl Mannheim, "Zur Problematik der Soziologie in Deutschland," *Neue Schweizer Rundschau* 36/37 (1929): 820–829.

26. Eduard Spranger, *Das deutsche Bildungsideal der Gegenwart in geschichtsphilosophischer Beleuchtung* (Leipzig: Quelle & Meyer, 1928), 2–3, 63 (hereafter cited as Das deutsche *Bildung*sideal); Ernst-Robert Curtius, *Deutscher Geist in Gefahr* (Stuttgart, Berlin: Deutsche Verlags-Anstalt, 1932), 14, 20, 44, 50 (hereafter cited as *Deutscher Geist in Gefahr*).

27. Curtius, *Deutscher Geist in Gefahr*, 63.

28. Spranger, Das deutsche *Bildung*sideal, 3 and 65.

29. Jürgen Oelkers, "Deutsche Klassik in deutscher Reformpädagogik. Übergänge um 1930," in Lothar Ehrlich/Jürgen John (eds.), Weimar 1930. *Politik und Kultur im Vorfeld der NS-Diktatur* (Wien, Köln and Weimar: Böhlau, 1998), 88.

30. Spranger, Das deutsche *Bildung*s ideal, 14–15, 54, 11.

31. For an invaluable record of the bitter impression that Spranger and his ideas made on an emigre intellectual when he presented them in Japan, as cultural promoter of the Axis alliance, see Löwith, *Mein Leben in Deutschland vor und nach 1933*, 112–115. The document gains special piquancy from the fact that Löwith's eventual position, after his final, painful break with a Heideggerean Nietzsche, resembles that of Spranger in Weimar.

32. Curtius, *Deutscher Geist in Gefahr*, 20, 42, 46, 56, 73–78, 120–121.

33. Curtius, "Sociology—and its limits," 117.

34. Karl Mannheim, "Letters from Heidelberg: Soul and Culture in Germany," in Karl Mannheim, *Karl Mannheim's Sociology as Political Education*, 79–97; Erich Kahler, *Der Beruf der Wissenschaft* (Berlin: Georg Bondi, 1920).

35. Edgar Salin, *Um Stefan George* (Godesberg: Küpper, 1948), 256.

36. Gerd Mattenklott, *Bilderdienst. Ästhetische Opposition bei Beardsley und George* (Frankfurt: Aisthesis, 1985).

37. Ludwig Thormaehlen, *Erinnerungen an Stefan George* (Hamburg: Hauswedell, 1962), 57.

CHAPTER TWO
THE LEGACY OF THE GEORGE CIRCLE

Ernst Osterkamp

How the idea of *Bildung* entered into the self-understanding of the German bourgeoisie before 1933 can not be seen better than in the circle formed around the charismatic poet Stefan George (1868–1922), as well as the other people influenced by him. The immense success of the natural sciences and the progressive industrialization around 1900 in Europe placed the humanistic concept of *Bildung* under strong pressure to legitimate itself anew. The inclusive concept of *Bildung* that had emerged in the eighteenth century, alongside the belief in the perfectibility of man, had to be newly defined around 1900, and redirected mainly toward the arts and human sciences. Criticisms of society or human life as increasingly mechanized or standardized by the market formed part of the *topoi* of the George Circle. Against the negative tendencies of modernity, the circle pitted an elite concept of *Bildung*, and at the core of this concept, for the George Circle, stood poetry, because poetry offered the highest expression of humanity's possibilities. Everyone belonging to the George Circle, and even those who were merely under the influence of George's poetry and writings, participated in the process of socialization to that cultured citizenry intended by the distinctive concept of *Bildungsbürgertum*. As Carola Groppe points out, moreover,

> The elite conceptions of *Bildung* of the Circle could assume a double function: they could, on the one hand, maintain an informal barrier for the *bildungsbürgerlichen* upper class against upwardly mobile groups, and, on the other hand, suggest via the reception of the writings of the Circle a potential mode of participation in a cultural discourse for a newly emerging *Bildungsbürgertum*.[1]

Not without justification does Carola Groppe define the "reconstitution of *Bildung*" as a central project of the George Circle.[2] Its members concentrated their activities on poetry, literary scholarship, and education, and in doing so contributed significantly to the academic fields and to the discourse of the middle class. Their influence thus extended far beyond the scope of academia. The life in and with *Bildung*, the reception of past and present culture, created a shared space of experience amongst the members of the elite group and the *bildungsbürgerlich* recipients.

This space was destroyed in 1933. In this year some of the most significant Georgians were forced into exile. The question arises whether and how, under the new conditions of exile, their concept of *Bildung* could be continued.

I begin by recalling the philosopher Karl Löwith who encountered members of the George Circle wherever he went in exile. In his recently published diaries, he records his exilic journeys from Rome to Sendai in Japan in 1936 and then from Japan to the United States in 1941. At one point we read of a meeting that occurred following his inaugural lecture at the Imperial Tohoku University on November 20, 1936 on "The Idea of Europe in German Philosophy of History." "After the lecture," Löwith wrote, "Herr K. Singer introduces himself to me, a German *Schicksalsgenosse* [sharer of our fate] who, apart from us, is the only German in Sendai. He visits us in our apartment and it pleases me to get to know this very clever and pleasant man, who largely shares my spiritual assumptions."[3]

Yet, despite this apparent commonality of spiritual assumptions, Löwith remarks in his next sentence that Kurt Singer was both a student of Georg Simmel and a disciple of Stefan George, neither of which was true of himself. Certainly Löwith never included himself amongst the disciples of the charismatic poet who lived from 1868 to 1933, even though, as a student, Löwith had had close ties with several members of the circle. George had assembled his group on the model of a master-apprentice relationship and Löwith understood this well. In his Curriculum Vitae of 1959 he noted that how, after the "breakdown of 1918" he had had to decide "whether to join the circle of George and Gundolf or follow, as a lone wolf, Heidegger, who," Löwith noted, "exerted no less a dictatorial control over the young spirits, albeit in a different manner." "I decided," Löwith wrote, "for Heidegger."[4] Löwith nevertheless remained in close contact with many members of the George Circle until 1933, and when he met Singer in Sendai in 1936 he was in a perfect position to understand both the nature of his thinking and his continued commitment to Stefan George in exile; and he was accordingly not unaware of the distance remaining between them.

Kurt Singer offers a unique opportunity for us to begin looking at the emigration of the George members. In 1931, Singer had been invited by the University of Tokyo to teach political economy, but he also brought George's idea of a "Secret Germany" with him to Tokyo, the idea of an elite state.[5] Having left Germany before 1933, he never experienced the breakdown of Germany's civilization for himself, and he was thus able to maintain a sharp distinction in exile between George's ideal of an elite Germany and Hitler's Third Reich. And George had, after all, died in 1933. From Tokyo, Singer saw nothing that suggested the decay of George's spiritual Reich. He remained blind to the ruptures emerging in George's circle between Jews and non-Jews or between the liberal moderates and conservative nationalists. For Singer, it was as if nothing had happened since 1931: he retained his imaginary vision of George's elite *Reich*, without realizing the disastrous consequences it would have for his political judgment. But Löwith saw this right away. Despite their shared "spiritual assumptions," then, Löwith saw how Singer's unabated elitism had led him straight into Fascism. "In politics," Löwith thus wrote of Singer, "he was a fascist. He hated all democratic institutions and even defended the Japanese invasion of China as a world-historical mission. [. . .] Nor did he want to be reminded of his Jewishness.

And rather than speaking of the plight of the Jews in Germany, he preferred to speak of the 'rape' of the Sudeten Germans by the Czechs."[6] Frozen in exile and cut off from reality, Singer represented, in Löwith's view, a Georgian orthodoxy, an idealized version of an ideological disaster. In his autobiography of 1940, Löwith named this idealized version "ennobled Nazism" [*Edelnazismus*][7]: a version in which George's spiritually elite state revealed itself by 1940 as the spiritual prefiguration of Hitler's state.

To complicate matters, Löwith did not think at first, based on his dealings with Singer, that this conception of genteel Nazism actually distinguished the non-Jewish from the Jewish members of the George Circle. But his opinion of this matter undoubtedly became more differentiated later, after he learned of the fates of the individual members—some dying in concentration camps, others being shot by the Nazis. This refinement of judgment may already have happened in 1941, after he arrived in San Francisco, when he visited one of the most prominent George members, Ernst Kantorowicz, on March 8, 1941, at the University of California at Berkeley. In 1927, under the George swastika, Kantorowicz had achieved instant fame with his monograph *Kaiser Friedrich der Zweite*. He was made a Professor in 1930 at the University of Frankfurt, but he lost that job, as a Jew, in 1934. He left Germany after the November progroms of 1938 and began teaching in Berkeley in 1939.

Like Kurt Singer, Kantorowicz remained a spirited supporter of George's "Secret Germany," even under the Nazis. But in what sense was he an supporter? What indeed, we may ask, happened to the Georgian orthodoxy under the sunny skies of California? Again, Löwith helps us out in his diaries, when he describes, with his usual sense of irony, how, one day: "I go to Ernst Kantorowicz, who has in the meantime secularized his George Reich, and who now says 'somehow.' A vast array of liquor in a grotesque movie diva apartment."[8] Apparently, then, under the sunny skies, George's state had diffused itself into a sybaritic relativism. In Löwith's view, Kantorowicz had become the stark antipode to Singer. Whereas Singer had held onto an old view to stabilize his identity in exile, Kantorowicz had thrown overboard George's holy teaching in order to join the new (American) culture of science. For Kantorowicz, exile provided a spiritual liberation from the hermeticism of the George Circle.

What we have so far are two quite different reactions in exile to the Georgian vision, the first of which suggests that the Georgian legacy was stabilized in a new dogmatism—which one would expect in a sect after the death of the savior; the second of which suggests that the Georgian legacy became diffused with the loss of the leader and the circle's ritual.

Of course, that there were these different reactions in exile is not altogether surprising. Nor should it be forgotten that there had always been tensions amongst the members, between those who believed in a purely elite community of the cultivated in the sense of the circle [*Bildungsgemeinschaft*] and those who saw the circle as the heart of a nationalist movement.[9] Such differences of conception were naturally highlighted in and after 1933. In part, they explain how the émigrés could continue the Georgian mission, how the Georgians responded to Hitler, and how the Nazis themselves reacted to the Georgians. Some Nazis would have liked to adopt Stefan George as a figurehead;[10] others saw him as a cosmopolitan bourgeois. What troubled all of the Nazis was the lack of racist conviction in George's vision.

Some of the Jewish members of the circle sided with the idea of the George state as a *Bildungsgemeinschaft*—Karl Wolfskehl, Edith Landmann, Ernst Gundolf, Erich von Kahler—but some sided with the nationalist alternative—Edgar Salin, Kurt Singer, Ernst Kantorowicz, Ernst Morwitz. Yet whatever their earlier views of George's "Secret Germany," the Jewish members of the circle all faced the same tragic situation after 1933. The Nazis' racism excluded them all with one stroke from George's "Secret Germany," in any sense, on the grounds of their not being German. Their leader and master, Stefan George could provide them no help or advice. He himself had left Germany in August 1933, and he had died in Minusio Switzerland, in December of the same year. In any case, his own attitude toward Judaism had always been ambivalent.[11]

And yet, the fact that he died in Switzerland turned out to be most relevant. Proclaimed the first exile of the George Circle, it would have been possible to present him as the leader who would lead all the Jewish members into exile. But this did not happen, for only three of the Jewish members actually left Germany in 1933. They were Wolfskehl, Landmann, and von Kahler. However, like George, they went to Switzerland. They joined a Swiss George colony that was most active in Basel. That Basel actually bordered on Germany was crucial, for it could be seen to represent a pure cultural—but still German—space divorced from Hitler's Germany. Here, the members could continue their project of a "Secret Germany" in some sort of political safety or in the pure or removed space of exile. Here, there was continuity in discontinuity, a cultural continuity under George's star that was made possible by the exile into a familiar bourgeois conservative German space, just on the other side of the border. If we want to know about the George legacy in exile, we thus need to keep in mind this factor, namely, that the establishment of a strong George faction in Switzerland before 1933 was what allowed the immediate continuity of the George Circle after 1933.

Thus, if George's influence was removed in Germany itself in and around 1933, it continued in Switzerland, as well as in Holland—which is to say, in places as close as possible to Germany, and thus in spaces that allowed the vision to remain at its most orthodox, esoteric, and epigonal. Before 1933, we may say, George's interpretation of the humanities had constituted a core of German culture; after 1933 it became both literally and metaphorically peripheral.

Apart from the early pioneers in Basel, the Jewish members of the circle resisted leaving Germany for as long as possible. They believed that they could uphold a separation between Hitler's Germany and their own esoteric conception of a true Germany, within Germany. Some Jewish members, notably Percy Gothein, in fact never left, and Gothein was in fact somehow able to work for the Resistance. Though Jewish, he died as a political prisoner in 1944, near Hamburg, in the Neuengamme concentration camp. Another Georgian, the most courageous Gertrude Kantorowicz, cousin of Ernst, died in 1945 in Theresienstadt. She had been the only woman of the circle to have her poems published in George's exclusive *Blätter für die Kunst*. In 1938, she helped to free another Georgian, Ernst Gundolf from Buchenwald.

Other Jewish members of the circle left only after the Pogroms of 1938: Ernst Kantorowicz, Ernst Morwitz, Ernst Gundolf, but also peripheral members such as Paul Friedländer. In 1938, Erich von Kahler left Switzerland to become an adjunct professor at Cornell. Karl Wolfskehl who had initially gone to Italy, now left Italy after "the Roman Caesar Monkey" Mussolini's visit to Hitler.[12] Wolfskehl went to New Zealand, to the point on the globe, as he wrote in a letter, as far away from Europe as possible on this little planet. Totally isolated, he died there in 1948.

Near or far away, none of the George members ever relinquished their commitment to the poet, Stefan George, even if their political adaptations of his vision varied. This charismatic figure of the poet continued to provide them an exemplary model for their own lives. Thus, in 1963, Kahler gave a talk about his "undying devotion to the poet." He had "to thank him," he said, "for something vital," namely "the knowledge that in the spiraling course of world events, one needs something on which to stand, and for which you stand, and that this something must be the form and dignity of man."[13] George as poet had offered a humanistic ethos, which no member, and no Jewish member, ever found they needed to relinquish. In George's poems, they found that common core that could provide them a continuity and commonality, despite, as I just said, their otherwise differentiated interpretations of the Georgian vision in exile.

Next, I summarize three rather ideal-typical ways of thinking about George's thought among the exiles. First, the continuation of the orthodoxy. Exemplary here was Singer with his double life as political economist and committed Georgian in exile. On the one hand, he extended his research on political economy to the culture of Japan; on the other hand, he maintained his Georgian world wherever he went—first in Japan, and later, as a lecturer in New South Wales in Australia from 1939 to 1957. Finally he went to his beloved Greece, the primal landscape of George's Platonic ideas, and remained there until his death in 1962.

Another example of this sterile Georgian orthodoxy is to be found in the case of Ernst Morwitz, who after 1930 was one of the leading judges in the Prussian State. For a long time, one of the closest friends of George, Morwitz was expelled from his job in 1935 and moved in 1938 to the United States, where he taught German literature at the University of North Carolina, Chapel Hill. In his exile writing, Morwitz devoted himself exclusively to writing about George, and he translated George's entire poetic corpus into English. On September 20, 1971, Morwitz even managed to die in Minusio, the place of George's death thirty-eight years earlier.

Second, the exilic movement of George's thought can be interpreted through the way that some members kept the idea of George as a poet, made him a pure poet by separating his poetry from his politics, and then began to think about a new Zionist politics based on their own Jewish origins. George's oldest friend, Karl Wolfskehl followed this path as did the philosopher Edith Landmann. Landmann left Germany in 1933 and wrote a twenty page essay in the same year entitled "Omaruru—to the German Jews who belonged to the Secret Germany." It has never been published. However, its basic utopian recommendation derived from a relentless analysis of the German situation, that George's "Secret Germany" ought to be established in Africa in view of the manifest impossibility of establishing it in Germany itself, it. "Let's be elsewhere in the world," she wrote, "what we were in Germany: Germans of Jewish descent, to whom the spirit of German poetry has been holy." Accordingly, Landmann does not think she will ever need another language, because, as she wrote "we [already] have the language of the poet."[14] But later, and increasingly conscious of the fact that she had created only a utopian vision, she moved her attention toward a real political alternative. Having earlier been hardly aware of her Jewish identity, she gradually found her way to Zionism. "I have become a Jew," she wrote.[15]

The third option concerns the adaptation of George's thought to the new thought the émigrés found in America. This adaptation basically consisted in removing the kind of Georgian anti-positivism that had led before 1933 to the mythified, dehistoricized, and monumentalized, so-called, *Gestalt* monographs. Only two members succeeded in really adapting their views, Erich von Kahler and Ernst Kantorowicz. Recall Löwith's ironic comment on Kantorowicz and the impact of the sunny skies on his liquor cabinet. But the adaptation had a more intellectual side too. His early "Gestalt" monograph, which I've already mentioned, on Frederick II was a masterpiece of monumentalized historiography.[16] Its literary strategy was such as to delete all contingencies in the Kaiser's life and spiritual existence. In every action, the Kaiser represented, according to Kantorowicz, a universalized leadership that pursued the destiny of the German nation. The mythifying of a dictatorial genius gave rise to a variety of political interpretation, especially in non-imperial times, and made this book a favorite of Hitler and Himmler. Goering sent a copy to Mussolini. It was precisely Kantorowicz's literary strategy, I want to emphasize, that allowed his book to become popular in a party that Kantorowicz himself despised.

Thirty years later, in 1957, he published his second masterpiece, "The King's Two Bodies." This book began with these words:

> Mysticism, when transposed from the warm twilight of myth and fiction to the cold searchlight of fact and reason, has usually little left to recommend itself. Its language, unless resounding within its own magic or mystic circle, will often appear poor and even slightly foolish [. . .]. Political mysticism in particular is exposed to the danger of losing its spell or becoming quite meaningless when taken out of its native surroundings, its time and its space.[17]

These words may be read as a subtle political self-criticism of the mythifying strategy of his earlier book. Still fascinated by the idea of political greatness and leadership, he now analyzed the symbolism of kingship quite apart from any identification with that power. In the first book, he had identified the worldly career of the Kaiser with the ideal form; in the new book he separated them. In the second book, he no longer stressed the myth of the monarch as such, but focused instead on the creation of an institution of monarchy by mortal men. In this late work, he was thus able to show methodologically, not the permanence but the mortality and changeability also of myth itself.

Given his own self-criticism, one can now understand why after the war he rejected the proposed republication of his first book in Germany. Only in 1963, just before his death, did he agree to have this book republished. But he did so only, he said, "to avoid giving the impression of a break with the legacy of the Circle [*Kreisvergangenheit*]."[18] Such a rationale, I think, was completely typical of the general posture of the Jewish German émigrés of the George Circle, for all of them maintained their fidelity to George, to his life and thought, without question. And, whichever of the three options they chose, whether they assumed the orthodox posture, or moved toward Zionism, or engaged in a scientific self-criticism of their earlier thought, they all proclaimed their loyalty to the poet and helped bring to an end the ideology of George's "Secret Germany."

Notes

1. Carola Groppe, *Die Macht der Bildung. Das deutsche Bürgertum und der George-Kreis 1890–1933* (The Power of Bildung: The German Middle-Class and the George-Circle, 1890–1933) (Köln, Weimar and Wien: Böhlau, 1997), 48–49 (hereafter cited as *Macht der Bildung*). The monumental work of Carola Groppe is of fundamental importance for situating the George-Circle within the history of *Bildung*. I thank Lydia Goehr for the translation of this article.
2. Ibid., 64. A comprehensive treatment of the life and work of George is provided by Robert E. Norton, *Secret Germany. Stefan George and His Circle* (Ithaca and London: Cornell University Press, 2002).
3. Karl Löwith, *Von Rom nach Sendai. Von Japan nach Amerika. Reisetagebuch 1936 und 1941* (from Rome to Sendai. From Japan to America. Travel Diaries 1936 and 1941), with an essay by Adolf Muschg, ed. Klaus Stichweh and Ulrich Bülow (Marbach: Deutsche Schillergesellschaft, 2001), 93 (hereafter cited as *Von Rom nach Sendai*).
4. Karl Löwith, *Mein Leben in Deutschland vor und nach 1933. Ein Bericht* (My Life in Germany. Before and after 1933) (Stuttgart: Metzler, 1986), 147–148 (hereafter cited as *Leben*).
5. On the ideology, politics, and science of the George Circle, I refer to three standard works, each of which contains extensive bibliographies. Klaus Landfried, *Stefan George. Politik des Unpolitischen* (Stefan George. Politics of the Unpolitical) (Heidelberg: Stiehm, 1975). Stefan Breuer, *Ästhetischer Fundamentalismus. Stefan George und der deutsche Antimodernismus* (Aesthetic Fundamentalism. Stefan George and German Anti-modernism) (Darmstadt: Primus 1995); Rainer Kolk, *Literarische Gruppenbildung. Am Beispiel des George-Kreises 1890–1945* (Literary Group-Formation. The Case of the George-Circle 1890–1945) (Tübingen: Niemeyer, 1998).
6. Löwith, *Leben*, 22.
7. Ibid., 20.
8. Löwith, *Von Rom nach Sendai*, 105.
9. These divergences are already visible among George's followers around 1910. They arise out of the conceptional differences between Friedrich Gundolf, who saw himself above all as historian of *Bildung*, and Friedrich Wolters, who sought to mount a political agitation for a new *Bildungsstaat*. Cf. Groppe, *Macht der Bildung*, 213–333.
10. The best treatment of attempts to turn Stefan George into an advertisement for the "Third Reich" is in Peter Hoffmann: *Claus Schenk Graf von Stauffenberg und seine Brüder* (Count von Stauffenberg and his Brothers) (Stuttgart: Deutsche Verlagsanstalt, 1992), 110ff.
11. On the reaction of the Jewish members of the circle on the political development in the year 1933 cf. Carola Groppe, "Widerstand oder Anpassung? Der George-Kreis und das Entscheidungsjahr 1933" (Resistance or Conformity? The George-Circle and the Year of Decision 1933), in Günther Rüther (ed.), *Literatur in der Diktatur. Schreiben im Nationalsozialismus und DDR-Sozialismus* (Literature in Dictatorship. Writing in the Era of National Socialism and GDR-Socialism) (Paderborn, München, Wien and Zürich: Schöningh, 1997), 60–92 (hereafter cited as *Widerstand*).
12. Karl Wolfskehl, *Briefwechsel aus Neuseeland 1938–1948* (Correspondence from New Zealand 1938–1948) 2 vols. (Darmstadt: Luchterhand, 1988), 926.
13. Erich von Kahler, *Stefan George. Grösse und Tragik* (Stefan George. Greatness and Tragedy) (Pfullingen: Neske, 1964), 31. On Erich Kahler, see the basic work by Gerhard Lauer, *Die verspätete Revolution: Erich von Kahler. Wissenschaftsgeschichte zwischen konservativer Revolution und Exil* (The Delayed Revolution. Erich Kahler. The History of Scientific Thought between Conservative Revolution and Exile) (Berlin, New York: de Gruyter, 1995).
14. Groppe, *Widerstand*, 91.

15. Michael Landmann, *Erinnerungen an Stefan George. Seine Freundschaft mit Julius und Edith Landmann* (Recollections of Stefan George. His Friendship with Julius and Edith Landsmann) (Amsterdam: Castrum-Peregrini, 1980), 137.

16. Ernst Kantorowicz, *Kaiser Friedrich der Zweite* (Emperor Frederick the second) (Berlin: Bondi, 1927).

17. Ernst H. Kantorowicz, *The King's Two Bodies. A Study in Mediaeval Political Theology. With a New Preface by William Chester Jordan* (Princeton: Princeton University Press, 1997), 3.

18. Gustav Seibt, "Die Unsterblichkeit ist eine Erfindung des Menschen. Das zweite Hauptwerk von Kantorowicz" (Immortality is a Human Invention. The Second Masterwork of Kantorowicz), *Frankfurter Allgemeine Zeitung* (25.3.1991): L 15.

CHAPTER THREE

WALTER BENJAMIN'S "SECRET GERMANY"

Irving Wohlfarth

Occult Enlightenment

Everything he said sounded as if it came from a mysterious place [Geheimnis]. *But its power derived from its self-evidence.*

—T. W. Adorno

"It is peculiar [*eigen*] to philosophical writing that at each turn of phrase it should confront anew the question of representation." Thus begins the "Epistemo-Critical Preface" to Walter Benjamin's study of German Baroque drama. "In its final form," it goes on,

> it will, to be sure, be doctrine. But it does not lie within the power of mere thought to impart to it such finality. [. . .] The part of method in philosophical projects is not reducible to their didactic organization. Which is to say that an esoteric element is peculiar to them that they are unable to discard, forbidden to deny, and that it would be their ruin to extol.[1]

Truth would thus have an inalienably alien element, an esoteric, mystico-theological dimension, as well as a teachable, transmissible, exoteric side.[2] This duality parallels Benjamin's definition of aura as "the unique appearance of a distance, however close it may be."[3] Such distance *appears* (to us), but it can never be *had* (by us). Truth is, in short, irreducible to knowledge. Whereas knowledge is "a having,"[4] truth cannot be owned. It exists *on* its own. "The death of intention,"[5] it is—*pace* Kant—a noumenal experience, or an experience of the noumenal, of "the death of intention,"[6] into which the subject enters and vanishes.

According to Hölderlin's poem *Mnemosyne*, truth "happens": "*es ereignet sich aber/ Das Wahre.*" It is "peculiar" (*eig(n)en*) to it that it should occur (*sich ereignen*) independently of our conscious will, beyond assertion or denial, taste or distaste. Such an event, or advent, is undreamt-of (or *only* dreamt-of) in purely academic discourses on method. No subject may appropriate (*sich aneignen*) or take credit for its appearance. Attempts to do so merely disqualify themselves. They repeat the Fall into subjectivity that brought down divine judgment on mankind's head.[7]

It was with the Fall, as Benjamin reads it, that the god-given Logos split into esoteric and exoteric parts.[8] To flaunt the esoteric or, conversely, to deny its existence thus mark symmetrically opposite ways of forgetting, and thus perpetuating, the Fall. Redemption can only lie in the healing of the split. On that Day, the esoteric and the exoteric will come together, or will at least have lost their mutual exclusiveness. Philosophy will no longer harbor a secret doctrine. It will have become an open secret, a teaching (*Lehre*) open to all. But it is a fond illusion to think that thinking could meanwhile presume to confer such completion on itself. Nothing ruins thought so much as its pretending not to be a ruin, its pretending to the role of the Messiah.

Hence the paradoxical superiority of partial genres such as the "esoteric essay" and the "philosophical project" (*Entwurf*) over the encyclopaedic philosophical systems dear to nineteenth-century empire builders hungry for the whole.[9] Unlike the closed system, these open forms anticipate the whole truth without already claiming to possess it.[10]

Hence too the superiority of a "dialectical" over a "Romantic" optic. "We penetrate the mystery (*Geheimnis*)," writes Benjamin in his essay on surrealism, "only to the degree that we recognize it in the everyday world, by virtue of a dialectical optic that perceives the everyday as impenetrable, the impenetrable as everyday."[11]

Or, as Benjamin puts it in the "Epistemo-Critical Preface, "truth is not the act of unveiling that annihilates all mystery but the revelation that does it justice."[12] The convergence of opposed terms—esoteric and exoteric, far and near, light and dark, enlightenment and revelation—momentarily collapses the hateful contraries of our postlapsarian thinking and prefigures the day when they will be completely ruined.

As so often, Benjamin is operating here between seemingly opposed camps— those who fish in the troubled waters of the occult (*im Drüben fischen*)[13] and those who claim that everything is perfectly clear. Clairvoyance is, in Benjamin's eyes, far too important to be left to professional clairvoyants or to their blinkered critics.[14] The hidden persistence of myth, magic and theology in an at once over- and under-enlightened era is not a professional secret that can be alternately withheld, advertised and capitalized upon. The occult is nothing to boast about. A more appropriate response to its continuing presence is one of shame—shame for an enlightenment that has so far produced as much obscurity as it has dissipated; shame, too, at its needing hidden leading-strings—once again *pace* Kant—and secret energies in order to guide it to the completion of its project.[15] The logic at work in the first of the *Theses on the Philosophy of History*, which makes the claim for such necessary secrecy, is that of the "Epistemo-Critical Preface." *Secrecy is maintained in order that it may one day finally be undone.*

An atmosphere of secrecy pervades everything that Benjamin wrote. As a "theme," however, secrecy leads a discreet existence in his work. This could hardly be otherwise: secrecy expounded is secrecy no longer.[16] Philosophical writing, our opening quotation suggests, has reasons that philosophical reason does not know. Benjamin's references to the esoteric are so many clues to a doctrine whose time has not yet come, intimations of the hollow ground on which all postlapsarian thinking stands.[17]

True, these hints had found undisguised theologico-philosophical formulation in Benjamin's early esoteric essays "On Language in General and the Language of

Man"[18] (which the hermetic "Epistemo-Critical Preface" to the *Trauerspiel* book again takes up) and "The Task of the Translator."[19] But he nevertheless treated the essay on language—which is in many respects the matrix of his thinking—as an *arcanum* and never sought to publish it. Likewise, Scholem, Adorno, and Missac all observed a characteristic secretiveness (*Geheimniskrämerei*) in his behavior. One may speculate on the biographical reasons for such excessive secrecy. It is clearly at odds with Benjamin's claim that discretion had changed from an aristocratic virtue into a petty-bourgeois vice and that "living in a glass house" was a revolutionary virtue par excellence.[20] It is more rewarding, however, to attend to the interplay between esoteric and exoteric forces that lies at the core of Benjamin's—and, if he is to be believed, of *all*—thinking. Thinking, reading, waiting, and loitering are all, he claims, so many forms of "profane illumination"[21]—a synecdoche unwittingly echoed by Brecht's dismissive jotting: "It's all mysticism—combined with an anti-mystical posture" (*Alles Mystik, bei einer Haltung gegen Mystik*).

Various overlapping areas of secrecy may be distinguished in Benjamin's writings. Three will be briefly isolated here: relations between Germans and Jews; those between modern Jews and their own tradition; and the operation of a Messianic force in human affairs.

"Secret Relations" between Germans and Jews

Benjamin's made his most important statement on German–Jewish relations in a letter he wrote in 1923 to a German-Protestant friend.[22] He is, he writes, forever conscious of his deep attachment to "the German" (*das Deutsche*), to which nothing binds him more deeply than his project of "saving" the Baroque *Trauerspiel*. But it is he goes on, equally clear to him that a German Jew cannot speak out on matters of national concern as if they were his own. He writes:

> Here if anywhere we are at the heart of the Jewish question in its present-day form: a Jew undermines even the best German cause to which he *publicly* commits himself, because such public German support is necessarily venal (in a deeper sense) and can produce no authentic credentials. Secret relations between Germans and Jews can, on the other hand, possess a wholly different kind of legitimacy. I firmly belief that all *visible activity* in the area of German–Jewish relations is bound to have disastrous effects and that a salutary complicity obliges the nobler representatives of both peoples to remain silent about their mutual bonds.[23]

Long after German Jews had been granted the civil rights that Werner Sombart had advised them not to exercise, their situation remained deeply compromised. Any claim on their part to speak on behalf of their country, Benjamin is here claiming, constitutes an unseemly act of bribery; the murder of the Jewish statesman Walter Rathenau should not therefore have come as a surprise. Honorable relations between Germans and German Jews can, in short, take place only in secret.

We might broadly reformulate this assessment as follows. Germany is itself a secret society that continues to refuse full membership to its Jews. The latter are thus left with little alternative but to perpetuate, in one form or another, the secret, clannish way of life that has always served their host societies as a pretext for suspecting and excluding them. Thus, German Jews can at present only be secret Jews and secret Germans. They are not entitled to intervene in national politics but only, for the time being, in cultural matters.[24] Benjamin's project of "saving" the Baroque *Trauerspiel* is a case in point.

Nor, Benjamin is arguing, should German Jews seek to resolve the tensions of their situation by either assimilating into Germanness or dissimilating into Judaism. He seeks rather to work from within his split heritage. His work as a politico-literary critic is born of the critical distance that this situation affords. Georg Simmel's and Gershom Scholem's respective notions of the inner "stranger" (*der Fremde*) and of "men from afar" (*Männer aus der Fremde*) overlap here with Marx's analysis of the socioeconomic alienation (*Entfremdung*) of an estranged pariah class and with various modernist aesthetics of estrangement (*Verfremdung*). The upshot is a "dialectical optic," which by slightly displacing its objects, places them in their "true," "surrealist," "Messianic," "revolutionary" light.[25]

Hidden Dealings between Modernity and Tradition

The "secret relations" between Germans and Jews were compounded by those that Scholem detected between certain exemplary modern Jews and the tradition from which they seemed to have been disinherited. The *locus classicus* of this acutely felt sense of loss is Kafka's *Letter to the Father*. And even though Freud disclaimed any allegiance to religion or nationhood, he admitted to "the secrecy (*Heimlichkeit*) of the same mental construction."[26] Obscure though he felt his attachment to Judaism to be, his essay *Das Unheimliche* (The Uncanny) had done much to elucidate the experience of what simultaneously did and didn't does and doesn't feel close to "home."

Benjamin and Scholem responded to this situation in overlapping but antithetical ways. Both saw it as their task to save an endangered past. Forced to emigrate from Berlin, Benjamin moved to Paris, where he embarked on his so-called Arcades project: a vast attempt to "awaken from"—and with—"the nineteenth century" and to recast the mass of motifs and images that had mysteriously reached him from hidden sources of Jewish mysticism into the exoteric mold of historical materialism. In a late letter to Scholem, he describes Kafka's work in terms of a similar ellipse between Jewish mysticism and the most brutal modernity.[27] A decade earlier, Scholem had voluntarily left Berlin for Jerusalem, where he began working out his own esoteric version of the Zionist project.[28] Combining this Jewish agenda with German philological methods and the Romantic notions of folk and nation that underlay them, his work on the Kabbala aimed to reconnect a reawakening people with the "secret treasures" (*geheime Güter*) of mystical tradition.[29] Far from being a spent force, Jewish mysticism had gone underground, and among contemporary writers Kafka and Benjamin bore witness to its continuing vitality. Its secrecy was, moreover, compounded by the clandestinity thrust upon it by the modern secular world.

It is this overdetermined secrecy that Benjamin allegorizes in the first of his *Theses on the Philosophy of History*.[30] Both here and in his Kafka essay, he interrelates the two poles of his German–Jewish existence by interpreting the "hunchback dwarf" (*das bucklicht Männlein*)—a figure from *German* folklore—as the fitting emblem for a *Messianic* theology and for Kafka's *Jewish* universe.[31] The dwarf (alias theology) hidden in a chess automat (alias historical materialism) emblematizes the subterranean link between German and Jewish folk traditions and the providential presence of a self-effacing mystical tradition—primarily, but not exclusively, Jewish in inspiration[32]—at the core of our seemingly transparent rationality. Here too a complex dialectic of esoteric and exoteric elements is in evidence; and the secrecy is compounded. While Benjamin's allegory admittedly airs the secret by disclosing the presence of the (un)canny dwarf, the secret agent's cover has not, it seems, thereby been blown. He is, at all events, shown to be successfully eluding the panoptic gaze of modern reason by disappearing into its machinery.[33] It is as if even the best, most militant enlightenment, left solely to its own devices, were in danger of becoming a useless automat or, worse yet, a totalitarian apparatus; as if its providential puppeteer, by occupying its blind spot, held out the promise of a more truly disenchanted Enlightenment.

The alternative may be put as follows. Scholem wants to *preserve* Jewish mysticism from the Enlightenment within which it has sought refuge; he accordingly will hold fast the possibility that it will one day reemerge. Benjamin wants to *blot out*[34] theology and hopes for the day when it will have disappeared into a more truly transparent apparatus. Till then the Enlightenment will remain an uncompleted project that needs to harbor a self-effacing secret.

The Secret Force at Work in History

The notion of secrecy makes two last appearances in Benjamin's *Theses on the Philosophy of History*:

> The past carries with it a secret index through which it is referred to redemption. [. . .] In which case, a secret rendez-vous (*Verabredung*) exists between past and present generations. Our coming was expected on earth. Like every generation before us, we have been given a *weak* Messianic power, to which the past has a claim. That claim cannot be settled cheaply. The historical materialist knows this.
>
> As flowers turn their head toward the sun, the past strives by virtue of a secret heliotropism to turn toward *that* sun which is rising in the sky of history. A historical materialist must be privy to this most inconspicuous of changes.[35]

A weak, imperceptible, but indestructible Messianic force links the "oppressed past"[36] with the endangered present. Each is thereby given the chance to "save" the other. This force is not a metaphorical trope but a metaphysical tropism—one that, however, operates only through human agency.

One instance of such a "secret rendez-vous" is Benjamin's version of a "Secret Germany." It marks a point of intersection between the three above-mentioned zones of secrecy. In turning this notion against its ideologues, Benjamin resituates it within the "secret relations" at work between past and present.

"Against a Masterpiece": Benjamin's Critique of Stefan George's and Max Kommerell's "Secret Germany"

For the mistake committed by a part of our youth [. . .] is that they seek in the professor something other than the man who stands before them—a leader (Führer), and not a teacher. *[. . .] the prophet after whom so many of the younger generation hanker is simply* not there *[. . .].*

—Max Weber, "Science as Vocation," 1919

Stefan George has been silent for years. Meanwhile we have acquired a new ear for his voice. We recognize it as a prophetic one. [. . .] What the prophet foresees are the courts of judgment (Strafgerichte).

—Walter Benjamin, "Stefan George in Retrospect," 1933

In the summer of 1929, Walter Benjamin wrote a review of a work of literary history entitled *Der Dichter als Führer in der deutschen Klassik. Klopstock. Herder, Goethe, Schiller, Jean Paul, Hölderlin* (The Poet as Leader in German Classicism), which had appeared one year earlier. The review was published in *Die literarische Welt* on August 15, 1930. On July 16 of the same year, the book's author, Max Kommerell, had (on the strength of another, more academic study on medieval metrics) received the nod from the same Johann Wolfgang Goethe University in Frankfurt that had withheld the same credentials from Benjamin five years before. The examining committee had given Benjamin to understand that his *Ursprung des deutschen Trauerspiels* (Origin of German Baroque Drama) was not acceptable as a *Habilitationsschrift*—the postdoctoral manuscript, which if approved, entitles its author to apply for a tenured position at a German university. On November 1, 1930, Kommerell gave his inaugural lecture as the holder of the post that Benjamin had failed to obtain.

It may be legitimately asked to what extent Benjamin's and Kommerell's contrasting academic fortunes may be traced to their differing choice of subject matter and method. Such suspicion, however, first needs to be properly stated. The chronology sketched above excludes the possibility that any petty rivalry entered into Benjamin's review. If competition there was, it did not take place at the psychological level. Not for nothing does Benjamin's review praise Kommerell for his "physiognomic, strictly non-psychological eye."[37]

What can, however, be shown is that Kommerell's book and Benjamin's review— and, correlatively, *The Poet as Leader in German Classicism* and *Origin of German Baroque Drama*—represent two conflicting versions of the "secret rendez-vous" that exists, according to Benjamin, between past and present generations. The rivalry lies, in short, between two opposing models of literary–political criticism. The critic is, according to Benjamin, "a strategist in the literary struggle (*Literaturkampf*)."[38] This was equally true on both sides of the divide; but the other side refused to admit it.

The review, entitled "Wider ein Meisterwerk" (Against a Masterwork),[39] perhaps marks the limit of what, at least in Benjamin's judgment, it was permissible at the time for a German Jew to say in public on a cultural matter. In his estimation, the book was indeed a masterpiece. Rarely had literary history been written this way. But while Kommerell's study abounded in astonishing insights, it was dangerously flawed by the ideology it served.[40] From Benjamin's perspective, it was a case study in the kind of esoteric discourse that does itself in. His final verdict resounds like the Last Judgment:

Not before this earth has been cleansed can it become Germany once again. Not, how-ever, in the name of Germany, and least of all that secret Germany that turns out to be the arsenal of the real, official Germany—an arsenal in which the vanishing cap (*Tarnkappe*) hangs side by side with the steel helmet.[41]

This judgment is crystal clear but dense. Whether or not it was thereby calculated to escape potential censorship is hard to say. It will, at all events, be necessary to supply some contextual information and to retrace Benjamin's argument in order to clarify what is at stake here.

Kommerell's first intellectual involvement was, like Benjamin's, with the "youth movement" (*Jugendbewegung*) and Gustav Wyneken. After separating from them, he joined another countercultural community—the circle of artists, academics, and intellectuals around the poet Stefan George. He soon became a protégé of Friedrich Wolters, the "politician" and later historian of the circle, and in 1924 succeeded Friedrich Gundolf as George's amanuensis, accompanying him on his travels for the next six years. "The master," as he was called, addressed Kommerell in turn as *Maxim*, "Puck" and "Kleinstes" (Tiniest one). It was during this "nine years' voluntary servitude (*Dienstbarkeit*)," as he later termed it,[42] that he wrote *Der Dichter als Führer in der deutschen Klassik*, which he submitted chapter by chapter to Wolters and George. Its aim was to establish a historical pedigree for the poet's—for example George's—claim to spiritual leadership. Brought out by the circle's publisher, Bondi, it was followed up a year later by a more poetic work written in a similar vein: a series of imaginary "Conversations from the Time of the German Reawakening" (*Gespräche aus der Zeit der deutschen Wiedergeburt*). George was, in short, the acknowledged *Meister* behind Kommerell's *Meisterwerk*.

In 1930, Kommerell parted ways with the master. His inaugural lecture in Frankfurt was devoted to Hugo von Hofmannsthal, and it closed by saluting in him a *Dichter* who was *not* a *Führer*.[43] Hofmannsthal had likewise moved to a safe distance from George.[44] He had also been greatly impressed by Benjamin's essay on Goethe's *Elective Affinities* (which denounces the ideology of the George school at work in Gundolf's book on Goethe),[45] and had published it in 1924–1925 in his journal *Neue deutsche Beiträge*.

The *George-Kreis* had all the characteristics of a secret society. It saw itself as the nucleus of a "Secret Germany" that was the harbinger of national renewal.[46] Nietzsche had written that a hundred men would suffice. "Unlike Nietzsche," Peter Gay observes, "George did not choose to be alone; it was the heart of his method to build a secret empire for the sake of the new Reich to come."[47]

Sects never regard history as an object of study, Benjamin observes, only as the object of their claims. Kommerell's book is, he claims, a case in point. It annexes classical Greece and German classicism to its cause. This tactic serves to hide George's own secret—the Romantic origins of his poetry—behind a classical facade.[48] "In effect," writes Benjamin,

this book lays the groundwork for an esoteric history of German literature with a radicalism unequalled by any of his predecessors from the circle. This is literary history only for the *profanum vulgus*; in reality, it is a sacred history (*Heilsgeschichte*) of the Germans. [. . .] Germany is the heir to the Greek mission: namely, the birth of the hero. Such Greekness is, needless to say, severed from all contexts and appears as a mythological force-field. [. . .] The author's chief concern is to construe classicism as the first canonic case of a German uprising against the times, a holy war of the Germans against the century, such as George was later to call for. Now it would be one thing to substantiate this thesis, a second to find out whether the struggle proved victorious, and a third to examine whether it was truly exemplary. These all blur together in the author's mind, but the third comes first. This means, however, that he considers struggle the paradigm, declares it to have been victorious and loses no sleep over his actual theme—the position of the respective parties.[49]

Nietzsche, one of the heroes of the George Circle, had designated himself an heir of the Greeks and had, before changing his mind about Wagner, called for a renewal of German culture. To that end, his polemical essay "On the Use and Abuse of History" invokes the need to find the right balance between "monumental," "antiquarian" and "critical" types of historiography, all of which, he claims, deny life if they go unchecked by one another.[50] In Benjamin's eyes, Kommerell, while combining all three modes, casts literary history in an all-too-monumental mold.[51] Benjamin responds by sketching a critical alternative.

How indeed did the parties stand? Is it permissible to reduce this complex process to the interplay between the heroic and the banal? Its complexion was oppressive, as the example of Goethe shows. There is plenty of heroism in the men of German classicism, but it itself was anything but heroic; its attitude was rather one of resignation. Only Goethe was able to maintain this stance to the last without buckling. Schiller and Herder went under; and those outside Weimar, notably Hölderlin, hid their heads from this "movement." Goethe's opposition to the age was, moreover, that of a restorative, dominating temperament (*Herrschernatur*), which had its source not in any classical past but in the bedrock of the most ancient power, indeed of the natural world at its most primeval (*dem Urgestein ältester Macht—ja ältester Naturverhältnisse selber*).[52]

Two versions of "myth" stand here in stark opposition. The one is neo-pagan and monumental, the other Jewish and critical. Kommerell weaves a Greco-German myth around Weimar classicism; Benjamin uncovers the mythic forces at work within it. All of Benjamin's work on Goethe is grounded in this insight into the all-too-mythic dimensions of a national myth.

Claiming to have no part in the petty present, the votaries of a "Secret Germany" vaguely look elsewhere. This is, in Benjamin's eyes, an aberration of a political, existential and historiographical order. Like Nietzsche before him, he is persuaded that historiography rests not on the allegedly unalterable facts of the past but on the most vital powers of the present. He will later term this shift of perspective the "Copernican turn in historiography." Here he puts it as follows:

The present may indeed be meager. Be this as it may, one has to have it firmly by the horns in order to interrogate the past. The present is the bull whose blood must fill the pit if the spirits of the departed are to appear at its edge. It is this deadly thrust of thought that is missing in the works of the [George] circle. Instead of sacrificing the present, they avoid it.[53]

"Why be a poet in a meager age?" (*Wozu Dichter in dürftiger Zeit?*), Hölderlin had asked. Meager though the age may be, Benjamin responds, it is the only one we have at our disposal. Only its energies can reawaken those of other times and places. Historians who avert their gaze from the present thereby fail to get their historical object into focus. They miss the "revolutionary chance" that is given to each present in the "struggle for the oppressed past." Far from being in secret agreement with such a past, the George Circle, as Benjamin here portrays it, reinforces the reigning literary canon and its own sphere of influence. In the name of a Secret Germany, it concludes a secret pact, a separate peace, with the powers that be.

Weimar classicism is, Benjamin observes, a "late and very statesmanlike discovery"[54] of the George school. Kommerell's *Dichter als Führer* seeks to combine Goethe's Olympian aloofness and Nietzsche's tragic *Pathos der Distanz* with George's posture of splendid isolation. The upshot is less an alternative *to* state power than an alternative version *of* it—a grand alliance between various claims to cultural leadership (*Führertum*) that are all rooted in the most ancient, mythical exercise of power.

The solidarity of the oppressed to which Benjamin appeals is thus counterbalanced here by that of the oppressors. "All rulers," Benjamin will write in the *Theses*, "are heirs to those who conquered before them."[55] While the notion of a *Führer* was one to which the young Benjamin had also adhered,[56] a commentary on its subsequent career may perhaps be heard in the resonances of an epitaph that Karl Kraus had written on himself and in the fact of Benjamin's quoting it in 1931: "*Zurück als Führer bleibt mein ganzes Irren*" (All my errors stay behind to lead).[57] Leadership and aberration are here both redefined in terms of one another.

At the risk of oversimplification, Benjamin's and Kommerell's conceptions of a "secret rendez-vous" between past and present may be contrasted as follows. The one stands under the sign of revolution, the other under that of myth, alias the "conservative revolution."

All of Benjamin's writings aim to "brush history against the grain."[58] The "spark of hope"[59] generated between far-flung acts of resistance interrupts the perennial continuum of domination. Just as fashion has a flair for the topical that actualizes the outdated fashions of another age, political historiography represents a "tiger's leap" into the past.[60] *Origin of German Baroque Drama* had, for example, unearthed the subversive potential contained in the seemingly "outmoded" trope of allegory.[61]

By contrast, the rendez-vous staged by Kommerell rests, as Benjamin reads it, on the mutual affinities between German Classicism, the George Circle and emerging Nazism.[62] Not unlike the latter, George's "Secret Germany" combines complicity with existing power with a semblance of rebellion against it.[63] Its secrecy is closer to that of the oppressors than to that of the oppressed. Its "reawakening" is anything but the "awakening" of the "dreaming collective" that will be the political program of Benjamin's Arcades project.[64] It is the reawakening of the German nation. *Deutschland, erwache!* will soon be the rallying call; and its "rise" stands in diametrical opposition to Benjamin's later invocation of the "sun that is rising in the sky of history"—namely, the revolution.[65]

It is now clear what Benjamin means when he calls the notion of a "Secret Germany" part of an "arsenal" in which "the *Tarnkappe* hangs side by side with the steel helmet." According to fairytale lore, a *Tarnkappe* is the magic cap with which goblins hide themselves from view. But its wearers are in this case diabolical counterparts to Benjamin's hunchback dwarf, and their camouflage (*Tarnung*) is all the more dangerous for seeming so harmless. It conceals them from friend and foe alike and doubtless also from themselves. They know not what they do; and this helps them do it. Their incognito masks a potential readiness to exchange a pose of esoteric opposition for open support of the rising new regime.

This, then, would be the all-too-profane secret behind "the Secret Germany." Benjamin decodes the term as a secret password to the corridors of power. It is, in effect, a call not for *another, better* Germany, but for the *same* one, transfigured and disfigured—the Germany that (Benjamin perhaps intuits) is about to hand over power to the *worst possible* Germany.

At its worst, Kommerell's book would thus belong to the category of *Edelfaschismus* (ennobled Fascism). The same year, Benjamin had entitled another review article "Theories of German Fascism."[66] Its subject had been the warrior rhetoric of essays written by the group around Ernst Jünger. One year later, he decodes the swollen terminology of a recent volume of essays in academic aesthetics as follows:

> The whole undertaking gives [. . .] the uncanny impression of a heavy-footed company of mercenaries that has come marching into the handsome, well-built house of art, ostensibly in order to admire its treasures—and at that moment it suddenly becomes clear that they do not give a damn for its order or inventory. They have moved in because of the positions it provides from which to shoot at [. . .] important defenses in the civil war. [. . .] the "beautiful," the "experiential" ("*das Erlebniswerte*"), the "ideal" and similar claptrap offer the best cover [. . .].[67]

The Jünger group, academic aesthetics, and the George school would thus these would be so many objective—and therefore secret—allies in the civil war that is being covertly waged in the arena of *belles lettres*. Three years prior to Hitler's seizure of power, Benjamin senses what the reckless call for a new, true "Germanness" and a "holy war of the Germans against the century" may soon turn out to have meant. In the "dangerous anachronism" of Kommerell's sectarian language (*Rune, Deute, Ewe, Blut, Geschick*) he reads the writing on the wall.[68] "If God ever struck a prophet with the fulfillment of his prophecy," he writes in 1933, "this was the case with George."[69]

"Not before this earth has been cleansed," the review of Kommerell's book concludes, "can it again become Germany. Nor can it be cleansed in the name of Germany, least of all in that of the secret Germany [. . .]."[70] It can be assumed that Benjamin is here deliberately citing a loaded metaphor against itself. He is presumably demanding a *cleansing of and from nationalist notions of cleansing*. The underlying assumption would then be that purification is, like clarity and clairvoyance, much far too important to be left to its impure ideologues. It is not, in Benjamin's eyes, hopelessly contaminated.

In other respects, too, Benjamin's and Kommerell's positions closely echo one another—sometimes, perhaps, too closely for comfort. Kommerell didn't, Benjamin writes,

> merely stick to past events: he also discovers what failed to happen. Note that he doesn't invent it, as if it were some phantasy-image, but quite simply discovers it as the truth that never took place (*der Wahrheit nach als ein Nichtgeschehenes*). His image of history emerges from the background of the possible, against which the contours of the real cast

their shadow. Likewise, nothing is highlighted for effect and the darkest, remotest areas are the best formed. [. . .] What sets this work and its secret intent apart [. . .] is the placing of its emphases. It would be absurd to look here for "historical justice." Something else is at stake.[71]

Benjamin's own conception of history may be described in remarkably analogous terms. Like Nietzsche and Kommerell, he rejects the historicist emphasis on "the way it really was" (Ranke's *wie es eigentlich gewesen*).[72] But he does so in the name of another idea of historical justice. Here too it is the placing of emphases that makes the difference.

This is, presumably, also true of the present essay. Benjamin would not here be portraying Kommerell as he actually was (and could not yet be in his own lifetime[73]) but in the light of his best and worst potential.

The above-quoted passage also suggests what is at stake for both sides. Their positions converge, diverge, compete, and there is perhaps not room for both. As in the case of the Romantic depreciation of allegory in the name of symbol denounced in the book on German baroque *Trauerspiel*, Kommerell's German ideology represents a neo-pagan perversion of the Messianic promise. This may partly explain the violence of Benjamin's response. It would be the perversion of his own position.

This violence points to another, more disturbing parallel between their respective stances. Benjamin claims to detect in Kommerell's language an unavowed complicity with "official" power. How, though, describe the power and the glory of his own rhetoric? In "The critique of violence," he had denounced "mythical" violence in the name of "divine" violence.[74] The question remains whether the line between clean and unclean violence can be so cleanly drawn. Is Benjamin's own language proof against the temptations of "unofficial," extraterritorial power? Was it perhaps in order to guard against this temptation that he described the Messianic force granted to each succeeding generation as "*weak*"?[75] Does the fact that it is pitted against the powers that be automatically place it beyond all possible entanglement with mythic power and guilt? Would not such a claim to immunity amount to another, Messianic alibi—the *Tarnkappe* of a hunchback dwarf? Benjamin spoke of Baudelaire's "metaphysics of the provocateur."[76] Should one speak in his case of a comparable "metaphysics of the stranger?"

In short, the question arises here whether, from Max Weber's point of view, Kommerell and Benjamin were both engaged in playing the prophet in a "godless [*gottfremden*], prophetless age."[77] This is not the place to discuss the potential inner dialogue between Weber and Benjamin or to explore to what extent Benjamin and Kommerell were "secret antipodes"[78] whose positions all too closely mirrored one another.[79] What initially needs to be shown is that Benjamin had his own version of a "Secret Germany."

Deutsche Menschen: Benjamin's "Secret Germany"

> For a secret Germany does indeed exist.
> —Walter Benjamin,
> *Gesammelte Schriften*, IV, 2, 945

Unable to earn a livelihood in Germany as a result of the restrictions imposed by the newly established *Reichsschrifttumskammer* on non-Aryan writers, Benjamin emigrated to Paris in 1933. Who was and was not a German was becoming a question of life and death. "Official" Germany was beginning to "purify" itself of its Jewish population.

One of the first books that Benjamin now hoped to complete was a collection of letters written between 1770 and 1870, which were intended to document what it had once meant to be a German. Like the so-called Arcades project for which he was now gathering materials, it combined eloquent quotation with sparing commentary.[80] These testimonies to a bygone world practically spoke for themselves. A "stranger" was here proposing to present-day Germans a "secret rendez-vous" with their own forgotten past. In an unpublished commentary, he put it as follows. At a time when the legacy of German humanism is being abandoned, these letters demonstrate the power that it once had to mold even private conduct.[81]

Between April 1931 and May 1932, Benjamin had published a number of these letters in the *Frankfurter Zeitung* under the pseudonym Detlev Holz. Given the worsening political situation, it subsequently proved impossible to find a German publisher for the entire collection. Eventually, a small edition was brought out under the same pseudonym in November 1936 by a minor Swiss press. Benjamin's hopes of smuggling a covert message to German readers were dashed. The book was destined to illustrate the story it told—that of radical German Enlightenment reduced to exile and obscurity.

The introductory commentaries contain a number of clues, Benjamin writes, "which, given how closely things are read over there nowadays, clearly hints at the real meaning of the title. I would at least like to think so."[82] This title—which he was willing to abandon for the sake of publishing the book in Germany—was as simple as it was eloquent: *Deutsche Menschen*. The writers of these letters were *Menschen* in the full Enlightenment sense: human beings whose nationality figured as a qualifying adjective. There could be no purification of Germany, the Kommerell review had concluded, "in the name of Germany." It could take place only in the name of a common humanity.

In the manuscript of a radio talk that would never be broadcast, Benjamin stated the aims of the collection by way of a quotation from the most prominent literary critic associated with the George school. In his all-too-monumental book on Goethe, Friedrich Gundolf had described the lives of the great artists in terms of three strata. Their conversations were the foot of the mountain, their letters the broad middle region, and their creative works the lonely, heroic peaks.[83] Benjamin proceeds to undo this hierarchy. The peaks of German classicism, he claims, are indeed "glaciated"—a frozen canon beyond discussion and without effect. Below the snowline, on the other hand, lies a zone of lively interchange between the great men of the age and their lesser contemporaries; and the testimony of their letters is all the truer for having eluded the wide-open jaws of the cultural apparatus (*Schul-, Presse- und Deklamationsbetrieb*). While certain correspondences have been incorporated into the canon or made to serve a hero-cult or, conversely, cited for their so-called human interest, this vast corpus of letters still mostly represents an unspoilt area of German letters.

Exclusion from the canon is thus not merely a misfortune but an opportunity. It makes possible a secret rendez-vous that testifies to the existence of a Secret Germany worlds apart from the nebulous and dangerous fictions invoked by the George school. This is how Benjamin put it in another unpublished preface:

> The purpose of this collection is [. . .] to show the face of a "secret Germany" that is nowadays usually enshrouded nowadays behind banks of murky fog. For a secret Germany does indeed exist. But its secrecy is not only the expression of its inwardness and depth. It is [. . .] the work of forces that have brutally and noisily denied it a public sphere of action and have thus condemned it to secrecy. It was the same forces that banned Georg Forster from his fatherland, forced Hölderlin to seek a living in France as a tutor, and played Seume into the hands of Hessian henchmen who sent him to America. Is it necessary to describe these forces in any further detail? Forster and Seume call them by their name, and in his most consummate poems Hölderlin invokes the figure of German genius against them. For none of these men—or the later ones to be included in this series—ever turned his creative work into an alibi for eluding the call of social need (*Aufruf der bürgerlichen Not*). It is because these letters make this so clear that they have remained so unknown. This is certainly true for Forster and Seume. And while Hölderlin's letters were read, what they have to say to Germans about Germany was what was least understood. There is some solace in all this: these letters have remained entirely untouched. Dinner and anniversary speakers have overlooked them. And while they have occasionally succeeded in falsifying these men's works—as if they had nothing to say, or rather no testimony to leave, to us—, a glance at their letters suffices to show where, now as then, that Germany stands that is still, alas, a secret one.[84]

Just as Benjamin had reclaimed the theological notions of symbol and allegory from Romantic usurpation in his *Origin of German Baroque Drama*,[85] he here saves the idea of a "Secret Germany" from neo-Romantic ideology. A Secret Germany indeed exists, he claims, but it is not the one that masquerades under that name. Its need for secrecy is not an ontological good but a historical evil. It is not here, as in the case of *das Geheime Deutschland*, a matter of interiority or mystery but of political repression, enforced clandestinity and the denial of a public voice. *This* Germany does not disdain the public sphere or exempt the "creative artist" from the needs of the day.

The forces that have reduced it to secrecy are, Benjamin claims, those with which the George Circle is secretly allied. From the late eighteenth century on, the "same" forces of oppression have been turning the best Germans into secret ones and expelling them from Germany. History contracts here into two opposing camps, each of which lays claim to its own Secret Germany.

There is one figure in whom both sides see the epitome of their Secret Germany: Friedrich Hölderlin.[86] Kommerell, Benjamin argues, had transformed Hölderlin's invocation of a community of spirit between Germany and Greece into a heroic memorial to the German future. "Over night," he goes on,

> ghostly hands will paint on it a large "Too late." Hölderlin was not one of those who rise from the dead, and the land to whose seers visions appear to over dead bodies was not his. Not before this earth has been cleansed [. . .].[87]

In *Deutsche Menschen*, Benjamin reclaims Hölderlin for a Greco-German community akin to the one he describes in his Kafka essay between German and Jewish

Volkstum. The poet, he writes in his commentary on a letter from Hölderlin to Böhlendorf,

> looks upon the countries which "the need of the heart and the need for food" open to his gaze as provinces of the Greek. Not the burgeoning ideal Greece but the withered real one, whose community of suffering with Western, and above all German, nationhood (*Volkstum*) is the secret of the historical [. . .] transsubstantiation of Greece (*Griechentum*) that is the concern of Hölderlin's last hymns.[88]

In the 1930s, German exiles and expatriates commonly claimed to represent the "other" Germany. *Deutsche Menschen* implicitly traces this Germany back to the best traditions of the German Enlightenment. So, admittedly, did almost all anti-Fascists, inasmuch as the Enlightenment was the overall target of Fascist attack. Benjamin's Enlightenment was, however, that of a radical minority that had been driven underground, banished, silenced or misread from the French Revolution onward and was now being systematically exterminated. Hence the paradox of an esoteric Enlightenment, in which secrecy is the ally of transparency, and reason the ally of mystery.

"The light of the public (*Öffentlichkeit*) darkens everything," claimed Heidegger,[89] who was known to initiates as "the secret king" of German philosophy. Ever since its inception, the Enlightenment has been accused of secreting darkness. The charge has been made from opposite sides—against the Enlightenment and on its behalf.[90] Benjamin's position may be roughly summed up as follows. There is no alternative to the light of the public sphere.[91] But as long as the forces of Enlightenment indeed darken the face of the earth, they will have to be secretly abetted by forces they thought to have behind them. They will need a hunchback behind their back. The dialectical image of a theological dwarf concealed in the seemingly transparent substructure of a materialist chess automat light is a plea for the strategic necessity of a crypto-Marxism in a new, publicly avowed sense.

Benjamin's "Secret Germany" communicated, in short, with the most radical traditions of the Enlightenment. On the day that the Enlightenment came into its own, relations between Germans and Jews would finally lose their secret malaise. The condition of the "stranger" would be superseded, or at least modified into that of a "foreign friend."[92] Till such time, however, the best Germans, Jewish, and non-Jewish alike, would have to remain strangers, foreigners, exiles. And so would all human beings worthy of the name. As the flood rose, *Deutsche Menschen* brought some historical specimens of a threatened species into an ark that included the collector in his collection.[93]

Notes

1. "[. . .] dass [den philosophischen Entwürfen] eine Esoterik eignet, die abzulegen sie nicht vermögen, die zu verleugnen ihnen untersagt ist, die zu rühmen sie richten würde," Walter Benjamin, *Gesammelte Schriften*, ed. Rolf Tiedemann and Hermann Schweppenhäuser (Frankfurt: Suhrkamp, 1974–1989), 1: 207 (hereafter cited as *GS*). *Richten* means both "to judge" and "to ruin." Cf. on God's *Gericht* and *richtendes Wort GS*, II,1,153.
2. Cf. on wisdom as "the epic side of truth" Walter Benjamin, *Illuminations*, ed. Hannah Arendt, trans. Harry Zohn (New York: Stocken, 1968), 87 (hereafter cited as *I*).

3. Benjamin, *I*, 222.
4. Benjamin, *GS*, I,1: 209.
5. Ibid., 216.
6. Ibid.
7. Cf. "On Language as Such and on the Language of Man," in *Reflections*, ed. Peter Demetz, trans. Edmund Jephcott (New York, London: Schocken, 1978), 326 (hereafter cited as *R*).
8. According to the "Epistemo-Critical Preface," there "inheres" [*eignet*] in post-lapsarian language, "beside its more or less hidden symbolic side, an open profane meaning." The task of philosophy is to reinstate the primacy of the symbolic dimension, which is the opposite of all outwardly directed communication (Benjamin, *GS*, I,1: 216–217). "Symbolic" is more or less synonymous here with "esoteric."
9. Ibid., 207.
10. Ibid.
11. Benjamin, *R*, 190–191, translation slightly modified.
12. Benjamin, *GS*, I,1: 211.
13. In a review of a book entitled "The Occult Sciences in the Light of our Time"— *Erleuchtung durch Dunkelmänner* ("Illumination from Obscurantists")—Benjamin cites the "melancholy question" of Karl Wolfskehl, a Jewish poet of the George Circle: "Should not the spiritists be said to fish in the beyond?" (Benjamin, *GS*, III: 357). *Im Drüben fischen*: a play on *im Trüben fischen*.
14. Cf. on clairvoyance "Madame Ariane—Second Courtyard on the Left" (Benjamin, *R*, 88–90).
15. Theology "today, as we know, is small and ugly and has to keep out of sight"; it is a "whispering campaign dealing with matters discredited and obsolete" (Benjamin, *I*, 253, 144, translations modified). Cf. also the notes to Benjamin's Kafka essay on the "indecency" of theology (Benjamin, *GS*, II, 3: 1233–1234).
16. Gershom Scholem's "Ten unhistorical propositions on the Kabbala" reflect on the irony of trying to subject secret doctrine to philological scrutiny, Scholem, *Judaica 3* (Frankfurt: Suhrkamp, 1987), 264–271.
17. In his essay "The Uncanny," Freud cites a passage from Karl Gutzkow, cited in turn by Sanders' German dictionary, on the sensation of walking on a dried-up pond or walled-in well: "We call that *unheimlich*; you call it *heimlich*," Siegmund Freund: *Gesammelte Werke* (Frankfurt: Fischer, 1961), XII: 234 (hereafter cited as *GW*).
18. Cf. Benjamin, *R*, 314–332.
19. The mystical doctrine behind these essays is expounded in an essay that Benjamin's closest friend spent half a century preparing to write. Cf. Gershom Scholem, "The Name of God and the Linguistic Theory of the Kabbala," in *Judaica 3* (Frankfurt: Suhrkamp, 1987), 7–70.
20. Benjamin, *R*, 180.
21. Ibid., 190.
22. Cf. on this issue my essays " 'Geheime Beziehungen.' Zur deutsch-jüdischen Spannung bei Walter Benjamin," special issue of *Studii Germanici* 28 (Rome 1990): 251–301 and "*Männer aus der Fremde*: Walter Benjamin and the 'German-Jewish Parnassus,' " *New German Critique* 70 (Winter 1997): 3–85.
23. Walter Benjamin, *Briefe*, ed. Gershom Scholem and Theodor W. Adorno (Frankfurt: Suhrkamp, 1966), I: 310 (hereafter cited as *B*).
24. In a much-discussed article entitled "German-Jewish Parnassus," Moritz Goldstein had gone one step further: "We Jews administer the intellectual propery of a people that denies us the right and the ability to do so." *Der Kunstwart* (March 1912).
25. Cf. Benjamin, *I*, 134.
26. Freud, *GW*, XVII: 52.
27. Benjamin, *I*, 141–145.

28. Cf. on this project my essay " 'Haarscharf an der Grenze zwischen Religion und Nihilismus.' Zum Begriff des Zimzum bei Gershom Scholem," in Gary Smith (ed.), *Gershom Scholem zwischen den Disziplinen* (Frankfurt: Suhrkamp, 1995), 176–256.

29. Hannah Arendt and Isaac Deutscher would, on the other hand, speak in entirely secular terms of the "hidden tradition" of the Jews as the "pariah" people of Europe.

30. Benjamin, *I*, 253.

31. After quoting the final verse of the folk-song, Benjamin writes: "In its depth, Kafka touches the ground which neither 'mythical divination' nor 'existential theology' supplied him with. It is the core of folk tradition (*Volkstum*), the German as well as the Jewish." (Benjamin, *I*, 134). At a time when Germans and Jews are being cast, by Nazis and Zionists alike, as mutually incompatible, Benjamin is pointing here to the common (under)ground between them. He is thereby reclaiming the contested ground of folk tradition from the Nazi myth of the *Volk* and perhaps also from Scholem's sometimes exclusionary Zionism.

32. Cf. on German Romantic Messianism: Benjamin, *GS*, I,1: 12.

33. Cf. on Jeremy Bentham's *Panopticon* Michel Foucault, *Discipline and Punish* (New York: Vintage Books, 1979).

34. Cf. on Benjamin's thinking as a "blotter" that would, if it could, blot out theology: Benjamin, *GS*, V, 1: 588.

35. Benjamin, *I*, 254, 255. Translation modified.

36. Ibid., 263.

37. Benjamin, *GS*, III, 253.

38. Ibid., *GS*, IV, 1: 108.

39. In letters to Scholem, Benjamin called the book "the most astonishing publication to have emerged from the George circle in recent years" (Benjamin, *B*, 1: 502) and described himself as having ruined his hands in George's garden on "the thorns of a rose-bush which, however, has in places some disconcertingly beautiful blooms" (ibid., 499).

40. In a review of Kommerell's *Jean Paul* written four years later, Benjamin would, however, claim that a certain distance already separated the earlier book both from its friends in the George Circle and from academic aesthetics (Benjamin, *GS*, III: 410).

41. "Nicht eher als gereinigt kann diese Erde wieder Deutschland werden und nicht im Namen Deutschlands gereinigt werden, geschweige denn des geheimen, das von dem offiziellen zuletzt nur das Arsenal ist, in welchem die Tarnkappe neben dem Stahlhelm hängt" (Benjamin, *GS*, III: 259).

42. Quoted in *Max Kommerell 1902–1944*, Marbacher Magazin 34 (1985), ed. Joachim Storck, 24 (hereafter cited as *Max Kommerell 1902–1944*.)

43. "To be a *Führer* means to solve the tasks of an age in ways that are valid for others. This was never Hofmannsthal's goal [. . .]. Where the victorious will fails, the pulse of the great riddles often becomes more audible." Cit. in *Max Kommerell 1902–1944*, 26. Cf. Kommerell's later diagnosis of George's will to power in his "Notizen zu George und Nietzsche," Max Kommerell, *Essays, Notizen, poetische Fragmente*, ed. Inge Jens (Olten, Freiburg: Walter, 1969), 226 passim. This move away from the *Führer* model also underlies the opposition that Kommerell sets up three years later between Jean Paul and Goethe: "He proved right in another way—not as a *Führer* but as a wise child or a holy old woman" (cit. Benjamin, *GS*, III: 417).

44. Cf. on this relationship Theodor W. Adorno, "George und Hofmannsthal. Zum Briefwechsel," in *Prismen: Gesammelte Schriften*, ed. Rolf Tiedemann (Frankfurt: Suhrkamp, 1974), 10,1: 195–237.

45. Cf. Benjamin's letter to Scholem of March 5, 1924 (Benjamin, *B*, I: 341).

46. George's poem *Geheimes Deutschland* speaks ambiguously of the gods having created, at the moment of greatest need, "new space in space" (*Neuen Raum in den Raum . . .*), and ends as follows: "Nur was im schützenden schlaf / Wo noch kein taster es spürt / Lang im tiefinnersten

schacht / Weihlicher erde noch ruht— / Wunder undeutbar für heut / Geschick wird des kommenden tages," Stefan George, "Das neue Reich," *Sämtliche Werke*. (Stuttgart: Klett-Cotta, 2001), IX: 46–49. The same collection contains a poem entitled *Einem jungen Führer im Ersten Weltkrieg* (To a Young Leader in the First World War), which consoles a handsome young warrior who had valiantly fought and lost (ibid., 31–33).

47. Ibid., 48.

48. "Every dialectical consideration of George's poetry will place Romanticism at its center; every orthodox, heroic account can do nothing cleverer than to minimize its role" (Benjamin, *GS*, III: 254). At the same time, it is, according to Benjamin, the German Romantic theory of criticism that constitutes the best answer to the George school's anti-critical, pseudo-Romantic espousal of creative innocence (ibid., 259).

49. Ibid., 254. What needs to be written, Benjamin elsewhere writes, is (not an esoteric history of literature but) a "history of esoteric literature" in which "the last page would have to show an X-ray picture of surrealism" (Benjamin, *R*, 184).

50. Benjamin likewise cites Florens Christian Rang, "the deepest critic of the German [*des Deutschtums*] since Nietzsche," on the morbid tendencies of Romantic idealism, the tradition in which the George school stands (Benjamin, *GS*, III: 254).

51. Cf. Max Weber's sober verdict on monumentality in 1919: "If we try to force matters and 'invent' a monumental aesthetic, the upshot will be a miserable failure, as with so many monuments of the last twenty years" (*Science as a vocation*).

52. Benjamin, *GS*, III: 256.

53. Ibid., 259. By contrast, Gert Mattenklott sees in Kommerell the singular case of a critic whose non-involvement with the literature of his own age in no way diminishes the quality of his work, "Max Kommerell. Versuch eines Porträts," *Über Max Kommerell. Zwei Vorträge* (Marburg: Universitäts-Bibliothek, 1986), 22.

54. Ibid., 254.

55. Benjamin, *I*, 256.

56. In letters written in 1913–1914, Benjamin evokes the good fortune of having grown up with a *Führer* such as Wyneken (Benjamin, *B*, I: 83. Cf. also 105–106 and note 5, 108) and speaks of the *Freideutsche Jugend* as a "circle" around student "leaders" such as himself (ibid., 109).

57. Benjamin, *R*, 273.

58. Benjamin, *I*, 257.

59. Ibid., 255.

60. Ibid., 261.

61. Cf. Benjamin, *GS*, I,1: 354; and on surrealism as "the first to perceive the revolutionary energies that appear in the 'outmoded.' " Benjamin, *R*, 181.

62. Cf. on the attitudes and fate of various members of the George Circle before and after 1933 Ernst Osterkamp's contribution to this volume; and on George's subsequent reception Michael Petrow, *Der Dichter als Führer? Zur Wirkung Stefan Georges im "Dritten Reich"* (Marburg: Tectum-Verlag, 1995). George himself died in self-chosen Swiss exile in 1933, "unwilling to lend his prestige to the triumphant Nazis whom he despised as ghastly caricatures of his elusive ideal" (Gay, *Weimar Culture*, 47). Benjamin's most complex statement on George is his review article *Rückblick auf Stefan George* (Benjamin, *GS*, III: 392–399).

63. Kommerell's account of the "poet as *Führer*" does not, however, anticipate the Nazi *Führer* cult—at least not in any simple sense. "The title of the book did not go uncontested even at a time when the concepts *Führer* and *Führertum* were current not merely among youth movement groups but also in the academic world. In a lecture given in 1924, Gustav Roethe applied the notion of the *Führer* to Goethe" (*Max Kommerell 1902–1944*, 16). Even George seems to have had doubts about Kommerell's title. He passes on to him without further comment the claim that it was "ill-advised simply to apply the ambiguous current notion of a 'Führer' to events from Goethe's age" (16). One has here an apparent

reversal of roles. It isn't George but Benjamin who finds Kommerell's introduction of the notion of *Führertum* into a discussion of Goethe's relation to Carl August, Napoleon, and Byron "more than instructive" (*GS*, III, 257).

64. Cf. *GSZ*, V, I: 490.
65. Benjamin, *I*, 257.
66. Benjamin, *GS*, III: 238–250.
67. Ibid., 287.
68. Benjamin, *GS*, III: 254. In 1930, Kommerell writes: "I read the first volume of Hitler's *Mein Kampf.* Narrow-minded, peasant-like, uncouth, but its instincts are mostly healthy and right." And in 1932 he was intently waiting to see whether the present dictatorship would develop into a "real *Führertum*" in which the Nazis would "coalesce, willy nilly, as a moved mass," quoted in *Max Kommerell: Briefe und Aufzeichnungen 1919–1944* (Olten: Walter, 1967, 27). From 1933 onward, however, Kommerell steadfastly rejected the Nazi régime.
69. Benjamin, *B*, 2: 578. Cf. similar remarks in 1940 (*B*, 2: 853).
70. Benjamin, *GS*, III: 259.
71. Ibid., 253.
72. Cf. Benjamin, *I*, 255.
73. "We do not yet know the fiery light with which history will illumine [George's] features on the day they acquire their expression for all eternity" (Benjamin, *GS*, III: 393).
74. Benjamin, *R*, 297.
75. Benjamin, *I*, 254.
76. Walter Benjamin, *Charles Baudelaire. A Lyric Poet In The Era Of High Capitalism*, trans. Harry Zohn (London: NLB, 1976), 14.
77. Cf. on Weber's complex relation to George, which combines admiration for his poetry with a critique of his circle as a charismatic sect, Arthur Mitzman, *The Iron Cage. An Historical Interpretation of Max Weber* (New Brunswick: Transaction Books, 1985).
78. Cf. *Max Kommerell 1902–1944*, 17.
79. One counter-indication will have to suffice here: Benjamin's short essay on Dostoievski's novel *The Idiot* (Benjamin, *GS*, II,1: 237–241). While Prince Myskin figures here as the self-effacing medium of the Russian nation, he is nevertheless the very opposite of a charismatic "leader." His attraction derives from the absence of any will to power. It is through this void that national energies are filtered into the promise of pure humanity. The final image of a volcanic crater recalls Hölderlin's *Der Tod des Empedokles*. Cf. my essay "The Birth of Revolution from the Spirit of Youth. Walter Benjamin's Reading of *The Idiot*," *Internationale Zeitschrift für Philosophie* 1 (1993): 143–172.
80. Adorno likewise compared these letters to Benjamin's memoir *Berliner Kindheit um Neunzehnhundert*. Both, he claims, unveil bourgeois self-concealment (Benjamin, *GS*, IV, 2: 949).
81. Benjamin, *GS*, IV, 2: 955–956.
82. Ibid., 2: 948. Letter to Max Horkheimer, 17.12.1936.
83. Ibid., 942.
84. Ibid., 2: 945–946.
85. *Benjamin, GS*, I, 1: 336.
86. At the beginning of the Hölderlin Renaissance stood the new edition of his poetry by Norbert von Hellingrath, who was close to George. Cf. on the ensuing cult Gay, *Weimar Culture*, 57–61.
87. Benjamin, *GS*, III: 259. Benjamin cites Kommerell as demanding of "patriotic literature" (*vaterländische Dichtung*) an "inner bondedness" (*das innigste Durchdrungensein von der Art des Stammes*) along with the "highest inner distance" and a proper sense of "shame" "Words," comments Benjamin, "that hint at the importance of the cultural resources here brought into play by the forces that are orchestrating a Germanic twilight of the gods" (ibid., 254).

88. Benjamin, *GS*, IV, 1: 171–172.
89. Cf. Benjamin, *I*, 31.
90. "Not the least of the tasks confronting thought today is to place all reactionary arguments against Western culture in the service of advancing enlightenment," Theodor W. Adorno, *Minima Moralia* (Frankfurt: Suhrkamp, 1962), 254.
91. Cf. Benjamin's characterization of George's "spiritual movement" as having "striven for the renewal of human life without considering that of public life" (Benjamin, *GS*, III: 394).
92. Cf. Benjamin, *B*,1: 96.
93. Cf. Benjamin's letter to Scholem of January 1937, quoted in Gershom Scholem, *Walter Benjamin: Die Geschichte einer Freundschaft* (Frankfurt: Suhrkamp, 1975), 252.

Chapter Four

A Humanist Program in Exile: Thomas Mann in Philosophical Correspondence with His Contemporaries

Reinhard Mehring

The Thomas-Mann Reception as a Representative Case

Bildung is the individual acquisition of an orienting knowledge. The "history of *Bildung*"[1] can be written according to its social *bearers*, legally constituted *institutions*, leading *ideas*, authoritative *founders* or *representatives*, or changes in *semantics*. The end of *Bildung*, according to Nietzsche, is the "sovereign individual" or "free spirit."[2] Thomas Mann followed him in this conception.[3] German neo-humanism was marked by the idea, influentially restated by Ernst Cassirer in 1916,[4] that "freedom" must give itself a "form." Mann accepted this, adding to it a Nietzschean concentration—also to be found in the George Circle—on the authoritative "founders" of a *Bildung*-form. He called them "the mighty ones," the "masters" of a nation, and he memorialized them in their "suffering" and "greatness." In his summative essay, "The Three Mighty Ones," he distinguished three "monumental figures"[5] of "the German genius," Luther, Goethe, and Bismarck, and he directed the nation toward Goethe. Mann viewed the history of *Bildung* as the history of the reception of the "mighty ones"—to use the language of Ernst Bertram—as the "legend" of the "myth" of a hero. Analogously, the history of the reception of Thomas Mann can be read for its representativeness in the history of German *Bildung*. In this, I do not limit myself to the historical findings, but rather hold fast to the philosophical question of Mann's validity as a German "master" of humane orientation.

One line of the philosophical reception of Thomas Mann accords with his self-understanding. It arose as an aspect of the discussions about humanist prototypes that were conducted in Germany during the "Crisis of Historicism," more intensely after 1918, and which were themselves exiled in 1933. This reception started out from the possibility of crediting Mann's humanist claims, and it viewed his work as one of the most important contemporary attempts to offer a relevant, morally and politically significant response to "the problem of humanity." At first sight, this topic seems to be only on the margin of exile studies, but when examined more closely, it becomes very significant to the question of transformation in German *Bildung*. Goethe and Mann stand for the beginning and the end of the "bourgeois epoch." In this tradition, Mann cast himself in the role of Goethe's representative. His defense of bourgeois *Bildung* and culture both before and after 1933 attracted a great deal of attention. Kurt Sontheimer has already emphasized the political significance of the topic, "Thomas Mann and the Germans."[6] More recently, there has been an even closer examination of the Mann reception as an indicator of German political culture.[7] The findings were predominately negative. Mann was much read, but intellectual criticism mostly distanced itself from him. The relationship between Mann and the German political mainstream was troubled.

To situate the significance of the topic more precisely and to see the special role of German–Jewish émigrés, it is first necessary to recall the position of bourgeois *Bildung* after 1900. What was already foreseen by the "new mythology" around 1800[8] became a widely shared conviction, at the latest after the "revolutionary breach" (Karl Löwith) of 1848 and the step in intellectual history from Hegel to Nietzsche: the assumption of a fall of idealist theology and philosophy, as well as an epochal change under the pressure of industrialization. The progressive de-christianization and secularization of life, not really mitigated by the organizational confessionalization during the nineteenth century,[9] came on the heels of the discrediting of the neo-humanist *Bildung*-world of the Goethian age. That is why the younger, youth-movement-inspired generation of the expressionistic war period held up Hölderlin against Goethe as the author with whom they identified. For figures like Thomas Mann, Ernst Cassirer, and even Eduard Spranger, Goethe became a code word in intellectual politics for the defense of the liberal democratic "bourgeois age." This was one motive for the pronounced Jewish reception of Goethe in the interwar years, except among outsiders like Walter Benjamin.[10] The philosophical reception of Mann by German–Jewish emigrants is also to be seen in this intellectual–political context.

These philosophic exile thinkers were actuated by the question of whether Mann's neo-humanist claim remained valid under the changed circumstances. It was not primarily an *aesthetic* discourse. Aesthetically, Mann was on the defensive. His place in the modern novel was contested. The avant-garde preferred Kafka. Nor was it a *social–historical* discourse. The change in the social foundations of German *Bildung* was, after all, apparent to all. It was above all a *political* and *philosophical* discourse. Its political meaning lay in the defense of cosmopolitan humanism against its nationalist detractors. Its philosophical meaning is not so easily grasped, but it is signaled by the interest of Cassirer—and later Hans Blumenberg—in Mann. Mann's thought belonged to the philosophical trend after Nietzsche. He independently addressed questions of contemporary philosophical anthropology, developing a grand overall conception in the philosophy of history, which comprehended the "humanization" of myth and a political–theological process of religious universalization, together with political centralization and pacification. Mann accepted the political consequences of these views, and stood up for them in an engaged and intergral way. This made him a representative of the "other" Germany, as it was proclaimed in the emigration.

As one considers this role today, one might expect that German literature specialists in the Federal Republic would have attached themselves to Mann as the most recent "classic" of their discipline. His work promised a new beginning in the neo-humanist tradition of German *Bildung*. Such a philosophical reception would have largely corresponded to Mann's understanding of himself. There were hardly any developed alternatives. To pick up the liberal democratic line of Weimar discourse and to build up Thomas Mann as the author to cite in order to demonstrate the humane effectiveness of German *Bildung* can be postulated as the logical course for these specialists to have taken. Yet that is not what happened. The (dis)continuity of German *Bildung* after the breach of National Socialism can thus be effectively investigated in the mirror of the reception of Thomas Mann after 1945 among scholars in the field of Germanistics. Did they renew the humanist *Bildung* discourse of the Weimar Republic? Did they make a new start in search of the "third humanism"? Did the study of Thomas Mann provide entree to such a new intellectual politics?

The academic subject of Germanistics had originated in the cultural prelude to 1848 as a national–liberal, political discipline. After 1900—between Scherer and Staiger—there was a new politicization and ideologization of the field, as well as—in response to Wilhelm Dilthey—the development of literary studies into a "cultural science." Karol Sauerland even speaks of a "paradigm shift under the auspices of philosophy."[11] Germanistics took an active part in the nationalist rejection of Versailles and the Weimar Republic. That is what divided Mann from the eminent Germanist, Ernst Bertram. Only recently has there been detailed research on the history of Germanistics and its double "semantic reconstruction" after 1933 and 1945.[12]

The reception of Mann is very interesting for this history because it was always political and contested. After 1945, these divergences became enveloped in the commonplaces of the Cold War. Mann's humanism was a major theme in the reception behind the Iron Curtain. The extensive research on Mann in the DDR continued the emigration's humanist line of interpretation. This could be shown by the case of Hans Mayer,[13] who published an important study of Mann, which also made an impact in the Federal Republic, especially among nonprofessional audiences.[14] The corresponding academic research on Mann in the Federal Republic, however, took different and new paths, which ignored Mann's humanist self-understanding and his politics. Aesthetically as well as politically, Mann did not appear to be especially important. Although Mann's works were widely read, he was hardly ever discussed as avant-gardist and humanist. This aesthetic indifference also manifested itself in a lack of popularity among writers.[15] Mann's politics were as contrary to the conservatism of the Adenauer years as to the hyper-politicized Left after 1968. Even Käte Hamburger, to be more closely considered below, turned away from Mann after 1945—and she played no more than a marginal role, in any case, in the Germanistics of the Federal Republic. The newer research at the time established itself at a remove from Mann's humanist self-understanding as well as apart from the mainstream of Germanistics, pursuing a de-ideologization and depoliticization of his work. The attempts at philosophical accreditation for which German–Jewish philosophers in emigration had pressed thus came to a dead end. Hardly noticed at the time by disciplinary research in literature in the West, it appears today as merely a remnant of Weimar discourse.

Those circumstances define my thesis. "Humanist *Bildung*" in Mann's sense barely existed after 1945, even as program. It seemed too closely tied to the social circumstances of the bourgeois world before 1918 and 1933 to be resuscitated. For Germanistics in the German Federal Republic, humanist literary receptions soon became mere history. This post-1945 failure to receive either Mann's humanist claim or its philosophical accreditation is a mosaic of the semantic rebuilding of practical discourse. Accordingly, my topic is first of all the philosophical reception of Thomas Mann, above all by German–Jewish authors in exile. In this, I limit myself to authors who were in personal contact with Mann and orient myself therefore by the correspondence as source. I begin with a brief review of Mann's humanist claim and then examine the cases of Siegfried March, Käte Hamburger, and Theodore W. Adorno. The scholarly interest is twofold, both historical and philosophical. On the one hand, I depict a disrupted line of reception, and, on the other, with normative and practical intentions, I point to the still present possibility of a philosophical reception and thus to Mann's humanism.

Mann's Breakthrough to a Humanist Self-Understanding

Terminologically, we can distinguish between humane contents and humanist self-understanding in Mann's works. Virtually any form of narrative literature makes up models of possible life. For this reason, artistic representations have humanizing effects, and an *aesthetic* humanism in art is to a certain extent tautological. Not every work of art, however, displays awareness of this and renders its humane contents explicit. Beginning, at the latest, with *The Magic Mountain*, Mann's work seeks to formulate and to justify humanist prototypes. It is very consciously involved in the literary exploration of possibilities and preconditions for one's own life. It is a form of "justification and rescue,"[16] as Mann put it, or, in a reminiscence of Baron Münchhausen, the hair by which "the Magician" attempts to pull himself out of the swamp of his perplexities, as Mann's individual response to his problem about the conduct of life.

Only with a clear distinction between morality and politics is it possible to conceive of a moral orientation of politics, which provides, in turn, the basis of a *political* humanism. Two aspects of his writings indicate both Mann's closeness to and his distance from contemporary academic debates on a "new," "third," and "political" humanism (W. Jaeger)[17] beyond historicism: his recurrent references to the Berlin philosopher, Ernst Troeltsch, and his disassociation from an one-sidedly cultural–scientific, historical–philological actualization of the *humanities*. During the 1920s, Mann was sympathetic to Ernst Troeltsch's concept of reconciliation. Taking his departure from insights into the contemporary misuse of philosophy of history as a wartime ideology, Troeltsch called, in a speech entitled "Natural Law and Humanity in World Politics," for a correction in the Romantic "counterrevolution" through the counterbalance of the universalist legacy, which, as Troeltsch expressed it, would elevate the Romantic idea of individuality to a "program of self-reflection in German historical-political-ethical thought," as well as to a demand for a new "cultural synthesis." Mann made numerous references to this essay.[18] Like Troeltsch,[19] he demanded a correction of the German Romantic *Sonderweg*, if also one that would, nevertheless, hold on to the significance of the Romantic idea of individuality for the future. In a later essay in 1936, Mann distanced himself from a one-sided, cultural-science restriction of the concept of *Bildung* to the "cultivation of the humanities" and he criticized the "bifurcation between *Wissenschaft* and *Bildung*."[20]

Mann consistently objects to the narrowing of *Bildung* to the sphere of the cultural sciences (*Kulturwissenschaften*), and he takes into account the contributions that both the natural sciences (*Naturwissenschaften*) and the experiences of nature could make to humane orientation. All this was already considered at length in *The Magic Mountain*. In that book, in the crucial chapter "Snow," Castorp draws the following conclusion from his "poetic dream of man": "For the sake of goodness and love, man shall let death have no dominion over his thoughts."[21] With its Kantian overtones, this sentence may be taken as the humanistic "foundation formula" of the work and thus, speaking epigrammatically, as the starting point of the philosophical reception of Mann. Philosophers like Käte Hamburger and Siegfried Marck took seriously the fact that *The Magic Mountain* not only formulates a morality, but that, in keeping with the Romantic model of the "absolute novel," it also attempts to provide this morality with a comprehensive discursive grounding.[22] Mann strove to take total and integral charge of the mediation and reflection of his own life in the public sphere; he also remained true to his humanist credo in practice through his commitment to the Weimar Republic. Among his contemporaries were many who understood this exemplary claim to the harmonious unity of life and work as an original philosophical impulse.

Philosophical Accreditation in Exile

The academy was not unaccustomed to the philosophical study of writers in the years before 1933. Nevertheless, an extensive philosophical concentration on a contemporary like Mann would have been an eccentricity in the usual career of an academic philosopher. In emigration, however, the career-paths of such academics were quite different than they had been before 1933. Preoccupation with Mann, who had received the Nobel Prize in 1929 and had a reputation as a sort of "king of the emigrants,"[23] held the promise of publication, sponsorship, and support. Most of the proponents of a philosophical interpretation of Mann's works also approached the "grand writer" (to use Musil's term) in person and solicited concrete assistance—in obtaining grants or in negotiations with publishers. Thomas Mann's response was as much professional as charitable. He thought unapologetically of literary criticism as an exchange transaction, grounded on mutual promotion. Accordingly, he also sought to influence American criticism through calculated self-interpretations,[24] as well as through his contacts to the prominent American newspaper publisher, Agnes Meyer.[25] And yet such elements of calculation do not detract from the integrity of Mann's moral and political commitments. The same can be said of the émigré philosophers. The fact that they were also pursuing personal interests as they struggled to survive under conditions of exile does not lessen the value of their efforts. The existence of a philosophical interpretation of Mann before 1933 in itself demonstrates the presence of objective grounds for their more extended study. If one inquires into humanist models, one does so without the intention of reducing those responsible for them to their social or political interests. Models are exemplary conceptions of practical maxims and rules of conduct. The humanist models constructed in exile are normative responses to a situation of extreme social and political crisis and danger. They cannot be considered apart from their historical, political, and biographical context, and ignoring the condition of emigration would be naive. On the contrary, the normative responses are especially interesting in relation to their concrete historical and political conditions. Yet, in the final analysis, they must still be discussed as normative and philosophical issues.

Mann was a prominent figure in the cultural life of Weimar Germany. He moved in a circle of authors, musicians, and literary critics. Philosophers as diverse as Georg Lukács, Alfred Baeumler, and Benedetto Croce sought his acquaintance. After 1933, then, Mann became a key figure of the emigration. A new group of philosophers, including Siegfried Marck, Ernst Bloch, Fritz Kaufmann, and Theodore W. Adorno approached him, in search of patronage or recognition. Yet they are also among the well-known philosophers who took Mann's claim to humanistic validity quite seriously. In the present study, I am concentrating on Siegfried Marck (1889–1957) and Käte Hamburger (1896–1992), as well as Theodor W. Adorno (1903–1969).

Siegfried Marck[26] gained his postdoctoral habilitation in Breslau in 1917, became an associate professor there in 1924, and held a chair in philosophy after 1930. He published a well-known, two-volume work on *Dialectics in Contemporary Philosophy* and was an active supporter of the German Social Democratic Party. In 1933, he sought asylum in France, and he became acquainted with Mann in Lugano at that time. He was able to emigrate to the United States in 1939, where he became a professor of philosophy in the charitable, part-time, mostly vocational Central YMCA College in Chicago, which he later helped to convert into the more widely recognized Roosevelt College. Marck remained in the United States after the war, but he visited the Federal Republic for lectures and, in 1955, taught for one year as visiting professor in Bonn.

In a polemical work published in Zürich in 1938, *Neo-Humanism as Political Philosophy*, Marck expressly depicted Mann's neo-humanism "from a philosophical point of view" as a "political and philosophical mission," in the context of a critical clash with both Fascism and Marxism, and he called it the "philosophy of the future." He attempted to formulate a philosophical program of a conservative revolution, as Mann had urged a year earlier, in the programmatic introduction to his journal, *Maß und Wert*. To this end, Marck began with an attack on what he saw as a convergence between the "Marxististic philosophy of world change" and the fascist transmutation of a conservatively revolutionary "myth of origin" into a progressivist "myth of anticipation." As positive counterforce, he represented Mann's neo-humanism as an "integral personalism," drawing mainly on Mann's conception of man as a "master of contradictions," as expressed in *The Magic Mountain* and elaborated in the essay on *Goethe and Tolstoy*. Marck laid out this thesis in its proximity to the philosophical anthropology of the time, and he called for a coalition of "humanist concentration" combining Christian, liberal, and socialist ideas and forces. In sum, he stressed the political significance of "neo-humanism," and, like Mann, he had the humanist "concentration" flow in the direction of a "militant," "socialist humanism." In this polemical treatise, however, he referred only to Mann's most basic principles, without offering readings of the literary works. Later, in an obituary, Marck sketched a project for a more rigorous classification of Mann as "artist-philosopher" in the tradition of Nietzsche.[27]

Mann did not approve of this monograph. In his journal, he noted: "Reading: *Zeitschrift für Sozialforschung.* Horkheimer, in opposition to Marck—who is in any case quite weak—assumes correctly that I am not comfortable with the philosophical–political message ascribed to me."[28] He evidently failed to recognize his own programmatic statements; and he was irritated by his cooptation to the "conservative revolution." And he was also annoyed by Marck's earlier "pleas for assistance."[29] Nevertheless, he remained in contact with Marck. The correspondence is as yet available only in extracts. Mann expresses himself in it with light irony. On September 19, 1941, for example, he writes to Marck:

> You must not forget that my political digressions are not composed, like your writings, from the point of view of absolute philosophy, but rather represent a kind of *haut* propaganda and possess a polemical–pedagogical character. We have had quite enough of German profundity, for the time being. [. . .]Goethe once said that the Germans should be prohibited for fifty years from uttering the word "feeling."[30]

Mann's interaction with Käte Hamburger was more extensive and complex, in keeping with the more richly detailed character of her studies of his work. Käte Hamburger was born in 1896 as the daughter of a Jewish banker in Hamburg;[31] she studied philosophy and literary criticism in Munich and received her doctorate in 1922 for work on "Schiller's Analysis of Man." She then moved on to Berlin and became a lecture-assistant to an associate professor of philosophy, Paul Hofmann.[32] Prior to 1933, she published works on Jean Paul and Novalis; and she also wrote a small monograph on "Thomas Mann and Romanticism," through which she came in contact with Mann as early as 1932. This "critical history"[33] imputes to Mann a distinctly Romantic way of experiencing the problem of the "idea of life," to which Mann himself could not give an adequate formulation. Two striking features of this study are its orientation to *The Magic Mountain* alone and its ambition of providing Mann with critical enlightenment about his own Romantic heritage. Hamburger sent the monograph directly to Mann. This initiated a lifelong correspondence, which deals with practical problems of emigration, but which also contains demanding and brilliant analyses of Mann's new literary productions. Adding vitality to the correspondence above all, however, is the promise and plan of a monograph on Mann's *Joseph* novel.

Soon enough, Hamburger is at work, expanding the genealogical approach of the 1932 study and attempting to situate *Joseph* in the *œuvre* as a whole.[34] The central question remains the conception of humanity in the figure of Joseph.[35] The ambitious project ultimately had to be cut back to a small work of *Introduction* with better publication prospects.[36] Following the appearance of the study, Mann expressed a positive opinion in public, if not in his journals.[37] Two years later, Mann was strongly vexed by Hamburger's "reserve towards, if not to say rejection of" *Doktor Faustus*.[38] The contact between them was maintained, but it ceased to have the excitement of a shared objective.

The preface to the introduction of 1945 refers to Mann as the "re-creator and re-inspirer of the idea of humanity emanating from the 'spirit of the age of Goethe.' "[39] The two principal chapters are called, "Joseph and the Myth, or The

Myth as a Reality of Life" and "The Myth and Joseph, or The Myth as Symbol." First, Hamburger examines how Mann views historical myth as a "reality of existence"[40] and as a "reality of life"; second, she shows how Mann gave this "historical novel" a humanist "symbolic significance."[41] In the end, then, Hamburger lays out the way in which Mann gradually raises the mythical veil to reveal the myth as symbol of the "secret" of humanity. The divine serves in the novel as expression and symbol of that "which makes man into man and distinguishes him from all other beings of the organic world, that he is a spiritual being that says I."[42]

Hamburger lived in Sweden for over twenty years. Not until 1956 did she return to Germany, when she was appointed an associate professor of literary studies in Stuttgart, where she published primarily poetological and philosophical studies. Her work as a whole has a philosophical character. Taking work of literary criticism as her point of departure, she worked out the poetological and ethical presuppositions of literary criticism. Her writings give voice to the view that literature is filled with ethical content and that it is capable of speaking the truth. These convictions were after all the basis of Hamburger's interest in Mann's conception of humanity.

From "A Bit of Actualized Utopia" to Philosophical Taboo: Theodore W. Adorno's Repudiation of Thomas Mann

Thomas Mann's relationship to Marck and Hamburger was on the whole a professional, strategic one. He valued them as propagandists of his work. The relationship between Mann and Adorno, is more difficult and multilayered. It is well known that Adorno, in American exile, was one of the most important interlocutors in the development of *Doctor Faustus*. Mann reports on this at length in his "novel" about *The Genesis of Dr. Faustus*. The correspondence, only recently published,[43] begins on October 5, 1943, with Mann's courting of the musical "connoisseur": "what belongs in my book must go into it, and it will also be absorbed by it." On June 3, 1945, on the occasion of Mann's seventieth birthday, Adorno writes an adulatory letter. Among other things, this includes:

> Who, it might be asked, has remained more true to the Utopia of youth, the dream of a world not disfigured by purposive calculation, than you, who have dedicated to it all of your ripeness and responsibility?[. . .]When I was privileged to meet you, here on this remote West Coast, I had the feeling that I was encountering in the flesh, for the first and only time, the German tradition from which I received everything—even the strength to withstand the tradition. This feeling, and the joy that it bestows— theologians would speak of a blessing—will never leave me. In the summer of 1921, in Kampen, I once followed you unobserved on a long walk and I thought to myself, how it would be if you would speak to me. That you have in deed spoken to me twenty years later is a bit of realized utopia that is hardly ever granted to anyone.

Even if some strategic hidden motives and a good dose of literary vanity enter in, this testimonial remains extravagant. Adorno joins his Utopianism so massively with Mann's person and works that he never again descends from this exalted level. In response, Mann very soon intensifies his courting with an insistent letter of

December 30, 1945. The talks with Adorno enter into *Doctor Faustus* in several ways, from the incorporation of musicological passages to a conceptual response. The great dispute of the novel over the "revocation" of Beethoven's Ninth Symphony, and thus of the humanistic promise of Weimar classicism as a whole, is not least a reply to Adorno. For Mann insists against Leverkühn and Adorno on the possibility of a revocation of "the revocation," just as he had earlier spoken of a "reaction against the reaction" of contemporary nationalistic irrationalism.[44] With Zeitblom, he defends the idea and possibility of the Beethoven work against the devil, and he disassociates himself from the idea of the medium of music as an articulation of lament. Against the musical lament, Mann wagers on the liberating force of the word that creates distance. His reflections on the relationship between music and literature, anticipated by Nietzsche's prelude, are not least a reply to Adorno.[45]

The correspondence becomes more varied after the appearance of *Doctor Faustus* and after Adorno's return to Frankfurt.[46] The novel and the well-known quarrel with Arnold Schoenberg are no longer at the center. The critical view of world affairs—the "German regression"[47] and "fascist America"[48]—comes to the fore. Adorno reports at length on German "tendencies of reaction" and the avidity of the student youth. "The comparison with a talmudic academy is irresistible. It sometimes seems to me as if the spirits of the murdered Jews have entered into the German intellectuals."[49] Against a background of apocalyptical expectations, they reassure one another of the coldness of the postwar war. "If it must be the end of the world, one wants at least to have been there."[50] They recall the experience of exile. Adorno calms Mann's uneasy thoughts of a "second exile" with the consideration "that the difference between America and Europe is infinitely *smaller* than our homesickness imagines."[51] What is really intriguing about the correspondence are the discordant shifts between Mann's humanistic view and Adorno's negativism. It is Mann who recalls Adorno to a critical view of Germany: "Ten horses could not drag me back to Germany."[52] "Germany is out of the question. For me, it is too uncanny."[53] And it is Adorno who softens his own negativism under the influence of Mann's person and work.

In view of the wretchedness of German postwar philosophy, he remains loyal to Mann's work and contemplates its "interpretation in a philosophical sense."[54] He presses repeatedly for the completion of *Felix Krull*. In this "myth of the nineteenth century,"[55] he invests the most "extravagant expectations"[56] and "unbounded confidence."[57] He hopes for an additional "bit of realized utopia" from this "child of fortune," since he maintains that he himself cannot do more than develop the "doctrine of right living" as a "dismal science."[58] Mann accepts this *Minima Moralia* with approval. In response to Adorno's *Essay on Wagner*, however, he remarked, "If only there were ever a positive word from you, respected sir, which granted a vision, however approximate, of the true society to be posited!"[59] Adorno replies:

> Now I do not want to drive the twilight of the idols so far that I fetishize the determinate negation itself; and while it is true that the force of the positive has gone over to the negative, it is no less true that the negation derives its right solely from the force of the positive. If this can be articulated at all in our time, as should certainly have to happen, or if asceticism has the last word, I simply cannot know, however much I have accustomed myself in my life to peer into the dark.[60]

When *The Black Swan* arrives instead of *Krull*, Adorno sees himself vindicated. Mann is close to Adorno in negativism; he interprets his last illness[61] in terms of a nature-philosophy that, as Adorno put it, "pursued the experience of the illusory character and vanity of life [. . .] to its materialistic conclusion."[62] A last deception between them is the hope of a reunion that runs through the correspondence and does not take the form of concrete planning until the last letter.[63] Just as Mann spoke of the "secret of the identity" between Zeitblom and Leverkühn, the correspondence proclaims the proximity between Mann's concept of humanity and Adorno's negative dialectic. It is a document of a congenial translation of humane hopes into a high-minded critique of the times. Perhaps nowhere else does Mann speak so pessimistically, and Adorno, with so much hope.

Adorno's 1945 reference to the encounter with Mann as "a bit of realized utopia" is not his only use of exalted language. On the occasion of a lecture by Mann on November 10, 1952 in the great hall of Frankfurt University, he composed a pane-gyrical welcoming address,[64] read by Horkheimer in slightly shortened form,[65] which celebrated Mann as representative of the tradition and "model" of "independence" and humanity. The address portrays Mann as Nietzsche's poet, as one who rescued the "Nietzsche of humanity" and gave form to the "truth contents of Nietsche's thoughts." This corresponded to Mann's self-understanding, since he had long contended that "Nietzsche did not find his artist as Schopenhauer had done, or at least not yet."[66] Mann wanted to be for Nietzsche what Wagner was for Schopenhauer. In 1952, then, Adorno is still in full agreement with Mann and his self-understanding. They are also closer in their politics at that moment than ever before. The welcoming address, which was not in fact delivered in his name, remained unpublished, however, until 1986. Adorno's letter about *The Black Swan*, which appeared in the *Neuen Rundschau* in 1962, seems, in contrast, pale and perfunctory. And the remarks, *Towards a Portrait of Thomas Mann*,[67] published that year in the same journal, hardly offer any encouragement to take seriously Mann's work in philosophy once more.

On the surface, the comments limit themselves to being of service in "countering some prejudices that stubbornly burden the person of the author."[68] They emphasize anew Mann's Nietzschean art of "disguise," oppose Mann's "objectivity" to the "myth of vanity," and note, "As much as he was lacking in vanity, he was no less given to coquetry."[69] The last word of the portrait is "Humanity." But the remarks are also aimed against "all the official artistic metaphysics in his texts": "Instead, I think the contents of an artistic work begin just where the intentions of the author cease. The latter are extin-guished in the contents."[70] While Adorno had still conceded in 1952 that Mann had poetically comprehended Nietzsche's "truth contents," he denied him in 1962 such access to his own intentions. In his *Aesthetic Theory*, he does not mince words:

> The philological procedure of getting hold of intention in the hope of capturing the contents is tautological and hence false: it retrieves from art works only what it stashed there in the first place. The secondary literature on Thomas Mann is one of the most unsavory examples.[71]

Adorno no longer has a candidly positive relationship to Mann's work. He does not put Mann on a level with Kafka or Schoenberg and he excludes him from the avant-garde. His aesthetic argument in this is the strict distinction between the "truth contents" of a work and the "self-understanding" of the author.

How is this change in the evaluation of Mann's works to be explained? Biographical reasons doubtless play a part. After Mann's death, Adorno heard ever more about the animosity felt for him in the Mann family, and even by "the Magician" himself. But this alone does not suffice as explanation. This change is much more likely due to a systemic imperative arising out of developments in Adorno's own work. With the development of his dialectical negativism, Adorno dogmatized his pessimistic philosophy of history to such an extent that even Mann's work, however much it might correspond to Adorno's negative Utopianism, was subject to the pictorial prohibition of hope and the forswearing of reconciliation. If this is accurate, then the short history of the philosophical ennoblement of Mann's humanism ends with Adorno's dogmatic repression of his original hopes in Mann. This can already be gleaned from the textual history—his marginalization of the earlier relations to Mann, as well as his failure to publish the "imaginary welcoming address" that postulated a picture of hope. Adorno's renunciation of utopia led to a dubious aesthetic theory, to a strict divorce of the "truth contents" of a work from authorial intentions. This theory dismissed the concord of "art and critic," for which Mann had sought all his life. Mann thus changed from bearer of hope to aesthetic antipode. The radicalization of the dialectical demand upon a work of art denied the author the chance for an aesthetic fulfillment of his ideals.

When we return to the original question about the representativeness of the Mann reception history for the history of German *Bildung*, we can take as our first finding the following: the end of the philosophical reception is revealed within the story itself, first, by Mann's ironic reservations and, second, by Hamburger's and Adorno's turning away from Mann. One may well suspect that the earlier attentions had exhausted their strategic significance. Hamburger and Adorno were no longer reliant on Mann's sponsorship and could dispense with strategic citations. Both paid him back for their earlier dependency. In Adorno's case, philosophical motives also played a part. The negativism of his philosophy of history increasingly precluded positive references. In any case, these findings do not support a strongly political interpretation of the motives for reception. It was not due to "the Germans" alone that Mann was no longer taken seriously in the Germanistics of the Federal Republic as an author with whom to identify and as representative figure of humanistic *Bildung*. He lost his prototypical function at that time among the emigrants as well. And yet the silencing of the philosophical reception continues to require explanation. Political conservatism was only a subordinate motive in the disciplinary reception. Rather, it was precisely due to the belated and forced reception of classical modernism and avant-garde that Mann came to be cast aesthetically aside. The judgment of Adorno had great weight. The West-German Mann reception was not unhappy about his turning away from the self-understanding of the author, since it made it legitimate for them to step out of the long shadow of Mann's self-understanding and to profess an independent understanding of the work, although in fact they mostly followed, even in their search for sources, the footsteps of the "Magician." Since the 1960s, Adorno's aesthetics has given Germanistics a proud self-confidence that it can discover the "truth contents" of a work apart from the author's intentions. The instrument of its depth hermeneutics has changed since the 1970s from the "dialectical" philosophy of history to historical–biographical psychoanalysis. Yet the strong theoretical justifications have ultimately been merely cover for a simple fact in the history of science, the "differentiation" of a historical–philological disciplinary perspective out of philosophical interpretative approaches.[72]

The scientific character of Germanistics, notwithstanding all of its theoretical armament, is indebted to nothing but hard philological findings. The most innocent explanation for the paradigm change within the Mann reception in Germanistics— its turning away from the philosophical self-understanding of the author and his early reception—then, would be the differentiation and emancipation of Germanistics from philosophical interpretive perspectives, as has already occurred in the history of science. The change in the Mann reception would then represent nothing but a paradigm change within Germanistics, its turning away from the humanistic *Bildung* assignment toward a strict model of *Wissenschaft*, and the representativeness of this change for German *Bildung* would remain an open question. The suspicion remains, however, that there is nevertheless a more direct connection between the disruption of the humanistic line of interpretation and the political culture of the Federal Republic. I do not want to say more about this.

On the Philosophical Yield of the Reception

The contributions of the authors discussed here to research on Mann are indisputable. Their aims, however, are not purely in the science of literature. These authors are united by the conviction that literature composes images of possible humanity. The work of Thomas Mann satisfies such interests in an outstanding way. Mann offers very concrete answers to the original ethical questions, already clearly posed in his early writings, as to man's possibilities for humanity in the world. And in his essays he also reflects on these questions categorically and philosophically. Mann's historical–political insights are also greater than is generally assumed. The man embodied, after all, an integral and exemplary unity of life and teachings in his public role. Quite apart from the quality of his answers, the search for a timely answer adapted to the present is in itself worthy of respect. Because the normative-practical questions are also posed here and now and because all "classically" inherited models give only limited answers, the importance of the philosophical reception of Thomas Mann as a facet of the humanistic discourse in exile should not be underestimated.

Notes

1. The literature on the semantics of *Bildung* and the history of *Bildung* in Germany is limitless. Cf. Rudolf Vierhaus, "Bildung," in *Geschichtliche Grundbegriffe. Historisches Lexikon zur politisch-sozialen Sprache in Deutschland* (The Rudiments of History: A Historical Lexicon of Socio-political Language in Germany) vol. I (Stuttgart: Klett-Cotta, 1972): 508–551 (which ends with Nietzsche and Lagarde's culturally pessimistic questioning of *Bildung*); *Handbuch der deutschen Bildungsgeschichte* (The History of German *Bildung*: A Handbook), ed. Karl-Ernst Jeismann and Peter Lundgreen. vol. III: 1800–1870 (Munich: Beck, 1987); *Handbuch der deutschen Bildungsgeschichte* (The History of German *Bildung*: A Handbook), ed. Christa Berg. vol. IV: 1870–1918 (Munich: Beck, 1991); Georg

Bollenbeck, *Bildung und Kultur* (*Bildung* and Culture) (Frankfurt: Insel, 1994); examples from the contemporary literature include: Eduard Spranger, *Das deutsche Bildungsideal der Gegenwart in geschichtsphilosophischer Beleuchtung* (The Contemporary German Ideal of *Bildung*: A Historical-philosophical Examination) (Leipzig: Teubner, 1928); Hermann Nohl, "Die pädagogische Bewegung in Deutschland" (The Pedagogical Movement in Germany), in Hermann Nohl and Ludwig Pallat (eds.), *Handbuch der Pädagogik* (Pedagogy: A Handbook). vol. I (Berlin: Beltz, 1933): 302–374.

2. Cf. Volker Gerhardt, *Friedrich Nietzsche* (Munich: Beck, 1992): 201–203.

3. Cf. Thomas Mann, *Gesammelte Werke in dreizehn Bänden* (Collected Works) (Frankfurt: Fischer, 1974), XII, 311 (hereafter cited as *Gesammelte Werke*). For the complete interpretation, refer to my monographs: *Thomas Mann: Künstler und Philosoph* (Thomas Mann: Artist and Philosopher) (Munich: Fink, 2001); *Das "Problem der Humanität."* *Thomas Manns politische Philosophie* (The "Problem of Humanity": Thomas Mann's Political Philosophy) (Paderborn: Mentis, 2003). This text summarizes in particular the reception-history studies of the newer book and situates them more specifically in the perspective of the history of *Bildung*.

4. Ernst Cassirer, *Studien zur deutschen Geistesgeschichte* (A Study of German Intellectual History) (Berlin: Cassirer, 1916).

5. Mann, *Gesammelte Werke*, X, 374.

6. Kurt Sontheimer, *Thomas Mann und die Deutschen* (Thomas Mann and the Germans). (Munich: Piper, 1961).

7. Thomas Goll, *Die Deutschen und Thomas Mann. Die Rezeption des Dichters in Abhängigkeit von der politischen Kultur Deutschlands 1898–1955* (The Germans and Thomas Mann: The Reception of the Author in Relation to the Political Culture of Germany from 1898 to 1955) (Baden-Baden: Nomos, 2000).

8. Cf. Manfred Frank, *Der kommende Gott* (The Coming God) (Frankfurt: Suhrkamp, 1983); Manfred Frank, *Gott im Exil* (God in Exile) (Frankfurt: Suhrkamp, 1988).

9. Cf. Franz Schnabel, *Deutsche Geschichte im 19. Jahrhundert* (German History in the 19th Century) vol. IV: *Die religiösen Kräfte* (The Forces of Religion) (Freiburg: Herder, 1937).

10. For an example of the generation-specific experience of Hölderlin cf. Hans-Georg Gadamer's impressive retrospectives in his own *Gesammelte Werke* (Collected Works). vol. 9 (Tübingen: Mohr-Siebeck, 1993).

11. Cf. Karol Sauerland, "Paradigmawechsel unter dem Zeichen der Philosophie" (The Paradigm Shift under the Auspices of Philosophy), in Christoph König and Eberhard Lämmert (eds.), *Literaturwissenschaft und Geistesgeschichte 1910–1925* (Literary Studies and Intellectual History) (Frankfurt: Fischer, 1993): 255–264.

12. An overview can be found in Jost Hermand's *Geschichte der Germanistik* (History of Germanistics) (Reinbek: Rowohlt, 1994); on the reduction of the "Humanities" to historical–philological literary studies, cf. an early work by Erich Rothacker, *Einleitung in die Geisteswissenschaften* (An Introduction to the Humanities) second edn. (Tübingen: Mohr-Siebeck, 1930); Heinz Schlaffer, *Poesie und Wissen. Die Entstehung des ästhetischen Bewusstseins und der philologischen Erkenntnis* (Poetry and Knowledge: The Emergence of Aesthetic Consciousness and Philological Cognition) (Frankfurt: Suhrkamp, 1990); for the resultant state of the subject cf. *Semantischer Umbau der Geisteswissenschaften nach 1933 und 1945* (The Semantic Reconstruction of the Humanities after 1933 and 1945), ed. Georg Bollenbeck and Christoph Knobloch (Heidelberg: Winter, 2001).

13. Cf. Hans Mayer, *Ein Deutscher auf Widerruf* (A German until Revocation) 2 vols. (Frankfurt: Suhrkamp, 1984/85).

14. Hans Mayer, *Thomas Mann. Werk und Entwicklung* (Work and Development) (Berlin: Aufbau, 1950).

15. Cf. *Was halten Sie von Thomas Mann. Achtzehn Autoren antworten* (What Do You Think of Thomas Mann? Eighteen Authors Respond), ed. Marcel Reich-Ranicki (Frankfurt: Fischer, 1986).

16. Mann, *Gesammelte Werke*, XI, 302.
17. Cf. my article *"Humanismus als "Politicum." "Werner Jaegers Problemgeschichte der griechischen "Paideia"* (The Humanities as "Politicum": Werner Jaeger's Critical History of the Greek "Paideia"), *Antike und Abendland* (Antiquity and the Occident) 55 (1999): 111–128.
18. Cf. Mann, *Gesammelte Werke*, XII, 627ff; XIII, 298–299, 597, 589.
19. Ernst Troeltsch, *Deutscher Geist und Westeuropa. Gesammelte kulturphilosophische Aufsätze und Reden* (The German Spirit and Western Europe: A Collection of Cultural-philosophical Essays and Speeches), ed. Hans Baron (Tübingen: Mohr-Siebeck, 1925), 3–27. In contrast to Mann's view, Troeltsch did not directly attribute the German idealism of Romanticism to Protestantism. Accordingly, in his research on history of religion, Troeltsch made sharp distinctions between the older form of humanism and the more modern humanist movement.
20. Mann, *Gesammelte Werke*, X, 346.
21. Ibid., III, 686.
22. Cf. Hermann J. Weigand, *The Magic Mountain. A Study of Thomas Mann's Novel* Der Zauberberg (Chapel Hill: The University of North Carolina Press, 1965).
23. For a somewhat dated overview, cf. Horst Möller, *Exodus der Kultur. Schriftsteller, Wissenschaftler und Künstler in der Emigration nach 1933* (A Cultural Exodus: Writers, Scientists and Artists in Emigration After 1933) (München: Beck, 1984); Anthony Heilbut, *Exiled in Paradise. German Refugee Artists and Intellectuals in America from the 1930s to the Present* (New York: The Viking Press, 1983). For a somewhat dated overview, cf. Horst Möller, *Exodus der Kultur. Schriftsteller, Wissenschaftler und Künstler in der Emigration nach 1933* (A Cultural Exodus: Writers, Scientists and Artists in Emigration After 1933) (München: Beck, 1984); Anthony Heilbut, *Exiled in Paradise. German Refugee Artists and Intellectuals in America from the 1930s to the Present* (New York: The Viking Press, 1983).
24. Cf. Wolfgang Adolphs, "Thomas Manns Einflussnahme auf die Rezeption seiner Werke in Amerika" (Thomas Mann's Influence on the Reception of His Work in America), *Deutsche Vierteljahrsschrift für Literaturwissenschaft und Geistesgeschichte* (German Quarterly for Literary and Intellectual History) 64 (1990): 560–582; cf. the following overview: Hans Wagener, "Thomas Mann in der amerikanischen Literaturkritik" (Thomas Mann in American Literary Criticism), in *Thomas Mann Handbuch* (The Thomas Mann Handbook), third rev. edn., ed. Helmut Koopmann (Stuttgart: Kröner, 2001): 925–940.
25. Cf. *Thomas Mann—Agnes E. Meyer. Briefwechsel 1937–1955* (Thomas Mann—Agnes E. Meyer: Correspondence 1937–1955), ed. Hans Rudolf Vaget (Frankfurt: Fischer, 1992), esp. the letter dated Sept. 7, 1941.
26. Siegfried Marck, *Der Neuhumanismus als politische Philosophie* (Neo-humanism as Political Philosophy) (Zürich: Verlag der Aufbruch, 1938); cf. Helmut Hirsch, "Biographisches zur Wiederentdeckung des Philosophen, Soziologen und Sozialisten" (The Biographical Rediscovery of Philosophers, Sociologists and Socialists), in *Ordnung und Theorie. Beiträge zur Geschichte der Soziologie in Deutschland* (Order and Theory: Contributions to the History of Sociology in Germany), ed. Sven Papcke (Darmstadt: Wissenschaftliche Buchgesellschaft, 1986), 368–385; "Thomas Mann und Siegfried Marck im US-Exil. Neues zur Biographie" (Thomas Mann and Siegfried Marck in Exile in the US: Biographical Additions), *Hefte der Deutschen Thomas-Mann-Gesellschaft* (Journal of the German Thomas Mann Society) 6/7 (1987): 70–86. Biographical data on such background figures is now available in Bruno Jahn, *Biographische Enzyklopädie deutschsprachiger Philosophen* (Biographical Encyclopedia of German-speaking Philosophers) (Munich: Saur, 2001); cf. Christian Tilitzki, *Die deutsche Universitätsphilosophie in der Weimarer Republik und im Dritten Reich* (The Philosophy of the German University in the Weimar Republic and the Third Reich) (Berlin: Akademie, 2002).

27. Siegfried Marck, "Thomas Mann als Denker" (Thomas Mann as Thinker), *Kant-Studien* (Kant Studies) 47 (1955): 225–233; cf. Siegfried Marck's "Thomas Mann als Dialektiker" (Thomas Mann as Dialectician), *Philosophie* 2 (1937): 112–138; Marck. (1954), 87–. Marck also wrote additional newspaper reviews on Mann's work.

28. Thomas Mann, *Tagebücher, 10 Bde* (Journals, 10 vols.) (Frankfurt: Fischer 1977–1995), 16.2.1939 (hereafter cited as *Tagebücher*).

29. Ibid., 12.11.1938.

30. Thomas Mann, *Briefe 1937–1947* (Thomas Mann: Letters, 1937–1947) ed. Erika Mann (Frankfurt: Fischer, 1963), 208.

31. This biographical information is taken from Hubert Brunträger's introduction to the correspondence with Mann. Gesa Dane provides a good portrait in "Käte Hamburger (1896–1992)," in *Wissenschaftsgeschichte der Germanistik in Portraits* (A History of the Field of Germanistics in Portraits), ed. Christoph König, Hans-Harald Müller, Werner Roecke (Berlin and New York: de Gruyter, 2000): 189–198.

32. On Hofmann and Hamburger's surroundings in Berlin, cf. Volker Gerhardt, Reinhard Mehring, Jana Rindert, *Berliner Geist. Eine Geschichte der Berliner Universitätsphilosophie bis 1946* (Berlin Geist: A History of the Philosophy of the University in Berlin) (Berlin: Akademie, 1999).

33. Käte Hamburger, *Thomas Mann und die Romantik. Eine problemgeschichtliche Studie* (Thomas Mann and Romanticism: A Critical History) (Berlin: Junker & Dünnhaupt, 1932); *Der Humor bei Thomas Mann. Zum Joseph-Roman* (Humor in Thomas Mann: On the Joseph-Novel) (Munich: Nymphenburger, 1965); *Über Thomas Mann. Der Briefwechsel Käte Hamburger—Klaus Schröter 1964–1990* (About Thomas Mann: The Correspondence of Käte Hamburger and Klaus Schröter 1964–1990) (Hamburg: Europäische Verlagsanstalt, 1994); *Thomas Mann/Käte Hamburger. Briefwechsel 1932–1955* (Thomas Mann/Käte Hamburger: Correspondence 1932–1955), ed. Hubert Brunträger (Frankfurt: Klostermann, 1999) (hereafter cited as *Briefwechsel*).

34. Mann/Hamburger, *Briefwechsel*, 23.12.1933.

35. Ibid., 26.4.1934.

36. Ibid., 23.1.1944.

37. Ibid., 31.1.1946.

38. Ibid., 2.2.1948.

39. Käte Hamburger, *Thomas Manns Roman "Joseph und seine Brüder": Eine Einführung* (Thomas Mann's novel "Joseph and his brothers": An Introduction) (Stockholm: Bermann-Fischer, 1945): 9.

40. Ibid., 81.

41. Ibid., 105.

42. Ibid., 126.

43. *Theodor W. Adorno/ Thomas Mann. Briefwechsel 1943–1955* (Theodor W. Adorno / Thomas Mann: Correspondence 1943–1955) ed. Christoph Gödde and Thomas Sprecher (Frankfurt: Suhrkamp, 2002) (hereafter cited as *Briefwechsel*).

44. Mann, *Gesammelte Werke*, XII, 244–245.

45. Due to impressions received at the conference held by Bard College, I have developed this argument, which appears here in greatly abbreviated form, in the ninth chapter of the 2003 monograph.

46. Cf. Marita Krauss, *Heimkehr in ein fremdes Land. Geschichte der Remigration nach 1945* (Return Home to a Foreign Country: The History of Return Emigration after 1945) (Munich: Beck, 2001).

47. Adorno/Mann, *Briefwechsel*, Adorno on June 6, 1950.

48. Ibid., Mann on March 8, 1954.

49. Ibid., 28.12.1949.

50. Ibid., 1.8.1950.

51. Ibid., 28.4.1952.
52. Ibid., 1.7.50.
53. Adorno/Mann, *Briefwechsel*, 19.4.1952.
54. Ibid., 1.8.1950.
55. Ibid., 3.5.1950.
56. Ibid., 2.1.1952.
57. Ibid., 28.4.1952.
58. Cf. Theodor W. Adorno, *Minima Moralia. Reflexionen aus dem beschädigten Leben* (Frankfurt: Suhrkamp, 1951), Zueignung; *Minima moralia. Reflections from damaged life.* Transl. E. F. N. Jephcott (London: New Left Books, 1974).
59. Adorno/Mann, *Briefwechsel*, 30.10.1952.
60. Ibid., 1.12.1952.
61. Ibid., 30.7.1955.
62. Ibid., 18.1.1954.
63. Ibid., 30.7.1955.
64. Theodor W. Adorno, "Imaginäre Begrüßung Thomas Manns" (Imaginary Welcoming Address for Thomas Mann), in Theodor W. Adorno, *Vermischte Schriften. Gesammelte Schriften* (Miscellaneous Works, Collected Works) (Frankfurt: Suhrkamp, 1987), XX.2: 467–472.
65. Cf. Max Horkheimer, *Gesammelte Schriften* (Collected Works) (Frankfurt: Fischer, 1975), XIII: 255–258.
66. Mann, *Gesammelte Werke*, XII, 84.
67. Theodor W. Adorno, "Zu einem Portrait Thomas Manns" (Towards a Portrait of Thomas Mann), in Theodor W. Adorno, *Noten zur Literatur, Gesammelte Schriften* vol. II: 335–344; *Notes to literature*, trans. Shierry Weber Nicolsen (New York: Columbia University Press, 1991).
68. Mann, *Gesammelte Werke*, II, 355.
69. Ibid., II, 342.
70. Ibid., II, 336.
71. Adorno, *Ästhetische Theorie, Gesammelte Schriften*, vol. VII: 226 (Aesthetic theory. Translated by C. Leonhardt (London, Boston: Routledge, 1984); *Aesthetic theory*. Newly translated, edited with a translator's introduction by Robert Hullot-Kentor (Minneapolis: University of Minnesota Press, 1997): 226; cf. the letters to Mann of July 6 and July 11, 1950.
72. Cf. Bärbel Rompeltien, *Germanistik als Wissenschaft. Zur Ausdifferenzierung und Integration einer Fachdisziplin* (Germanistics as a Science: Towards a Differentiation and Integration of an Academic Discipline) (Opladen: Westdeutscher, 1994).

Chapter Five

The Empire's Watermark: Erich Kahler and Exile

Gerhard Lauer

The "long nineteenth century" came to an end in 1914, and with it the validity of its concepts of *Wissen* and *Bildung*. For the intellectuals of the time, such an ending appeared fated, since it was a standard gesture of cultural criticism in the empire to predict the inevitable collapse of the unprecedented heaping up of knowledge. Nietzsche was neither the first nor was he alone in his belief that *Bildung* and culture were in a state of crisis. In the second part of his *Thoughts out of Season* he passes in review, to great effect, the prevalent critique of the accumulations of merely antiquarian knowledge, remote from life. Others would follow. In 1911, for instance, Georg Simmel made out a "tragedy of culture," contending that the nineteenth century had piled up a "stock of spiritual objectifications that reached to the sky,"[1] and that such stocks could no longer be restored to living subjectivity. *Bildung* had lost all meaning. What had once been the purpose of *Bildung*, deepening an understanding of oneself and one's own culture, had simply ceased to exist. The diagnosis included all areas of *Wissenschaft* and art, even society itself. Society had produced a knowledge that was of no use to any *Bildung*, and that had become quite alien to it.

The younger generation born of the empire was not satisfied with a mere critique. Movements were called into being, with the "overcoming of the nineteenth century" as program. The *Yearbook of the Spiritual Movement* was one such programmatic manifestation. Its authors were close to the circle surrounding the charismatic poet Stefan George, many of them his disciples.

One of those whose voice was heard in those years was Erich Kahler, not a member of the inner circle around George, but in his sphere of influence. He appeared in the *Yearbook of the Spiritual Movement* of 1912, with a critique of the theater. The state of the theater, he maintained, mirrored the spirit of the age, which possessed more *Bildung* than any earlier time. Yet this *Bildung* is mere amusement. [. . .]The state of affairs in the theater is no different than that in the *Wissenschaften*, or—as Kahler calls it—the "*Wissenschaft* hothouse."[2] The crisis embraces everything. Only a few, like the circles around Stefan George, are exempt. They must be the "leaders" out of the crisis.

Every such critique of *Bildung* portrayed itself as exclusive, but by 1912 the arguments had long become common property.[3] In 1888, in "The Case of Wagner," Nietzsche had issued the watchword: "That theatre may not become overlord of the arts."[4] Now it seems to have become the master of modern culture, Kahler concludes, and the fall of the world of the fathers is therefore ineluctable, in 1888, as in 1912 and 1914—when it is overdue. In this crisis of European culture, a *new man* must be born. To become new, he would have to ripen apart from society, as in the intellectual counterworlds[5] of Heidelberg or Berlin. "The way into the innermost, most natural community," Kahler promised, "leads in these times inexorably away from the external and bourgeois one. That this is as always the only way to grasp life in its true form, in its own place, the others will not admit, and they deceitfully rework their weakness into other principles."[6] The Stefan George Circle, however, was a "new community" that sought to find ways into life, and their leader on this one and path toward cultivating the *new man* was the poet, George, who was at the same time the symbol of what was needful. Within his person the *new man* is already taking shape. He was also the exemplary model for Kahler.

With distance, it is not difficult to recognize the extent to which this idea of life and its *Bildung* represents a heightening of the leading ideas of nineteenth-century middle-class *Bildung*. The question of *Bildung* is now elevated to one of ethics and character, as the objectifications of *Bildung*, its institutions, move into the background. It is life that must be unconditionally grasped. This is reiterated to the point of monotony. Yet what this life was supposed to be remained unclear. At issue was more the reverence with which the teachings of the great spirits were encountered than the teachings themselves. *Bildung* retained the aura that had elevated the arts and literature above the quotidian of the bourgeois world, but it was now an aura that no longer knew its center. Kahler wanted not only to share in this aura, but also to be himself the new man and poet, and to achieve this with the unconditionality and decisiveness that inspired the expressionist pathos at the time of World War I, an ethos that required "full sacrifice of the person, sustained inwardness and great outward self-denial, and moral exertion,"[7] as Kahler himself expressed it.

The constellation of the *new man*'s unmistakeable origins in the cultured bourgeoisie together with his antibourgeois pathos was not atypical of the period, and its basis in Kahler's biography are readily identified. Kahler was born in 1885 as a Jew and a German into the *haut bourgeoisie* of Prague. His childhood and youth followed the picture book path of the rich and gifted *jeunesse d'orée* of the Hofmannsthals, Rathenaus, Warburgs, and Manns. Kahler's mother, Antoinette von Kahler, was a frequently published author of children's books; his father, Rudolf, had risen to the highest management positions in the insurance business, first in Prague and then in Vienna. The aristocratic "von" was an outward indication that the Kahlers lived at the heart of society. Although this is commonly called "assimilation," nothing could be a less accurate characterization of the historical circumstance of this social transition. The Kahlers and the other families, whether in Prague or Vienna, had not adapted themselves to any existing high bourgeois society, but had themselves materially contributed to its formation and design. Erich's mother's expectation of raising "young men of significance"[8] is borne by the self-certainty of belonging to the cultural elite. Pride in their own bourgeois modernism was unbroken, and not only in the Kahler family. There was no question that Erich Kahler would study philosophy and history, as well as the histories of art and literature, that he would be in a position to become a writer, or that his best-beloved cousin, Eugen von Kahler, who died young, would become a painter in the *Blauer Reiter* circle. Convinced of their own cultural mission, they had all the evidence on their side.

Kahler's "own ways" from which "life" was to be learned, which he invoked in a juvenile poem, were symbolically staged through a breach with his father. If Erich Kahler's father was a man of industry, representative of the insurance and financial enterprises that had only come into being in the nineteenth century, his son was a poet. His father's world was that of the Habsburg Monarchy. Kahler had no love for it; he had longed for the fall of the multinational state. That it was this world, which had raised him up and which had shaped his humanism more completely than any programmatic proclamations was not and could not be a subject of discussion. All this belonged to the expressionist revolt, which wanted to reinvent mankind as if it had never had a history. In 1919, Kahler's book on the *Habsburg Kindred* had appeared and caused a great stir, capturing the mood of such different authors as Egon Friedell and Hugo von Hofmannsthal. History had moved beyond the world of the fathers, he wrote. The liberal bourgeoisie, no less than the "thoroughly thoughtless and irresponsible House of Habsburg,"[9] had slithered into the war, which had made their loss of legitimacy manifest. Because the old world possessed neither life nor youth, it was "irretrievably dead and gone."[10] It consisted of historical reminiscences and neo-baroque longing, according to Kahler, but these are now nothing but façades of *Bildung* and they have ceased to be life-forming energies. In Kahler's cultural–philosophical overview of the old Kakania, Hofmannsthal had found "certain of his own reflections on the Austrian mentality [. . .] tellingly expressed here,"[11] and it was not coincidental that he compared Kahler's book, *The Habsburg Kindred*, to Oswald Spengler's *The Decline of the West*. For both, European civilization had outlived itself. Both promulgated a tragic vision of its downfall, and foretold an "Age of a New Man."

For Kahler there was a locale for this world-historical decision, and that was Germany. The second volume of the Hapsburg book, never actually written, was supposed to demonstrate "the compelling necessity, for the renewal of German Austria, of an intensive injection of the refreshing popular (*volkhaft*) spirit of the German Reich,"[12] as it was portentously put in the publisher's prospectus. With some show, Kahler put off his aristocratic title and moved to Germany. "Quite early, urged by some need for more bracing air and a less casual attitude to life, he came to the Reich,"[13] Thomas Mann writes, looking back on Kahlers decision. In his *Doktor Faustus-Novel*, Thomas Mann later created a memorial of his friend, Kahler, in the figure of Konrad Deutschlin. As so often in Mann's work, the name "Deutschlin" speaks for itself. It testifies to Kahler's high-minded belief in Germany, although Mann did not deny himself a certain distanced irony in his depiction of his Deutschlin character. For Kahler's pathos when the talk was of Germany, in exile no less than before, put off even Thomas Mann. In a diary entry of 28 September, 1935, Mann claims that Kahler, "goes dangerously far in his objective acknowledgment of the positive elements in what is happening (not being done) in Germany."[14]

At the distance of a quarter of a century, Thomas Mann took note of that radical turn amongst intellectuals around the time of World War I, but he also recorded their attempts to find a way out of such extremism. Totalistic terms like "life" merely reiterated these emphases. They gave no indication, however, as to what this new, more humane life should consist of. Kahler was certain of only one thing: that a breakthrough was imminent and that it would be equivalent to a revolution. "Revolution is here, whether we want it or not," Kahler asserted between the wars. "The thing is to want it. The thing is to let the order come into being and not to obstruct it by 'putting affairs in the old order.' Only when there is an end to the call for order, when the call is for the life that is coming, only then will there be order."[15] Such paradoxical figures of thought, which combined corporatist ideas of an order of estates with revolution, have been called "conservative revolution."[16] "So the new is essentially also the arch-human,"[17] was a saying of a leading figures of the popular university movement, which he sent to Kahler with a friendly dedication. The emerging new humanism consisted of formulas, each outbidding the other. Its demand was to replace the differentiated resources of *Bildung* with a unity of life orientation. For Kahler, as for Krieck and other intellectuals of the time, this presupposed that the accumulations of knowledge could be compounded and transformed into a power to revolutionize life as a whole. This was the new task of *Bildung*, which was thought no longer able to rely on the inherited institutions of transmitting knowledge, or to require objects of *Bildung*. In dispute were not the contents of this new *Bildung*, but rather the ways—or "the leader"—toward the goal.

From the standpoint of nineteenth-century historicism, it would have been impossible to compound knowledge into a force for *Bildung*. Modern society was already at that time generating a body of knowledge of which it could not say which parts might at some time become relevant. The new *Wissenschaft*, in contrast, which sought to revolutionize life overnight, had to conceive of knowledge as directly functional, which was incompatible with the "nonpurposive," anti-teleological modern forms of knowledge. For its new departure, then, the younger generation was left with nothing except a return to the philosophy of history from the German idealistic tradition. First they transmuted modern accumulations of knowledge into a *Bildung* that gave the appearance of being immediately present to "life," a theme endlessly repeated, and not only by Kahler. For anyone who argues from philosophy of history has the world spirit on his side. He knows that the world of the fathers "inevitably" had to meet its end. He can explain why the new is so "irresistible." With this, *Bildung* in the sense of a bourgeois form of life oriented to art, history, and the *Wissenschaften*, as well as its institutions, like the Gymnasium and University, is suspended and superceded. The antibourgeois circle around George, the popular university movement or the movement for a new life all promoted a new *Bildung* for man, beyond the established institutions. This new *Bildung* no longer required subject matter, but only a state of mind. *Bildung* is elevated to an ethos—but one empty of contents.

What this means can also be well followed in what was doubtless the most widely discussed of the writings of Kahler, *The Vocation of the Sciences*. The title already discloses it as a polemic against Max Weber's lecture, *Science as a Vocation*, in which Weber had postulated that only heroic persistence in an irretrievably modern and demystified world offers a chance for humanistic self-assertion. The *Wissenschaft* to be pursued was "value-free," which meant that its findings were not unmediatedly capable of providing orientation to life. Everything is different in Kahler's writing. It promised an escape from the entombment of modernity. *Wissenschaft* would lead to a "supra-rationality." The discrete domains of knowledge were to be joined together and to turn into a teaching for life. This conjunction was to be the task of the "leader and teacher." "The leader and teacher we have in mind, who is above all a human being in the highest sense and who only attains to his knowledge and practice through and out of the human being, this leader and teacher must prove himself wherever he goes or stands: in the tone of his voice and in the look in his eyes, as in the form of his works and the example of his conduct."[18] Here, *Bildung* had become an ethos wholly without subject matter. The distinctive disciplines with their respective subject matters are only of secondary interest. What mattered was the magnification of the acquisitions of *Bildung* into a meta-science. The German idealist tradition calls this "*Wissenschaftslehre*" (doctrine), while Kahler calls it *Beruf* (vocation); but how it was supposed to take place remained unclear. All that was clear was that it had to happen, and that the task had been given unto only a few. The calling, not the *Wissenschaft*, was of prime interest. The new *Wissenschaft* would necessarily follow:

> The succession of ever more dense because ever more detailed causal nets, the ever more numerous array of causal chains, the ever more diversified disordering of empirical reality through countless numbers of conceptual rays—all this has helped to initiate a new fluidity and cyclicality (!), an organic, irrational, wholly internal closure in all dimension, which manifests itself with ever greater density and clarity the more the conceptual and mechanically causal dissections strain to keep up.[19]

Ernst Troeltsch's judgement of Kahler and of the many similar writings of the years around World War I was that they were "basically 'revolutionary books against the revolution.' "[20] They are revolutionary because of their promise of a *new man*, and thereby also of an exit from the "demystified world." They are counterrevolutionary, because their struggle is against what Kahler calls the "old *Wissenschaft*," hallmarked by "concept," "abstraction," the "Kantian deed," "mechanism," "specialization," "intellectualization"—all of them things that the new *Wissenschaft* will no longer require. This *Wissenschaft* had no method, and needed none. In a modern sense, it was no longer *Wissenschaft*. What Kahler and many of his contemporaries in Germany and Austria had in view was the end of modernity, with which they could never develop an easy relationship. From its end, they promised themselves a new Humanism; and Germany was to be the land of this restoration.

What came in 1933 was quite different than expected, and took a long time to accept. Kahler was not prepared for the ferocity that persecuted him in 1933 and searched his house in Wolfratshausen near Munich. Intellectuals like Ernst Bertram or Ernst Krieck, who Kahler had thought shared his convictions, changed sides to the new powers. Kahler's books and essays lost their readership overnight. Thanks to his wife, Josefine, and to his family's good relations with President Benesch, Kahler was able to escape, by way of Vienna, Prague and Zurich, and finally, in 1938—at the urgings of Thomas Mann and his family—to Princeton. Kahler was forty-eight years of age when he had to abandon Gerrmany, the land of destiny. He had more than half his life behind him. The coming of the *new man* seemed more remote than ever.

It has been argued, especially in exile studies—spurred by the emigrants themselves—that exile contributed to a deprovincipalizing of the German tradition of thought, as Paul Tillich had already maintained.[21] I do not believe that this applies to Kahler, first, because the internationalization of the sciences had already begun before 1933, and, second because the world of ideas of a Kahler suddenly found itself without resonance, which is how it would remain. The playing field of thought became smaller. A new direction became more difficult. America had for too long appeared to incarnate everything about modernity that one had rejected.

Even from an external point of view, Kahler was unable to secure lasting employment, to say nothing of a permanent professorship. Grants from the Van-Leer-Foundation, temporary teaching at the New School (1939/40) and at Black Mountain College in North Carolina (1946/47), as well as a number of visiting professorships at Cornell University (1947–1955), Ohio State University (1956 and 1959), and Princeton (from 1960) and other universities provided his academic daily bread, and his audiences consisted mostly of undergraduates. It was not until shortly before his death that he was recognized with an honorary doctorate by Princeton University. Where could a thinker like Kahler build a link in United States? There were of course first of all the emigrants themselves. His fraternal friend Hermann Broch lived for several years at Number One, Evelyn Place, Kahler's house in Princeton, and Kantorowicz, Einstein, and Wolfgang Pauli were not far away. But they served rather to reinforce one another in their own intellectual aristocratism, which still waited for a radically new departure in the world. In this expectation, they kept too much to themselves. Wolfgang Pauli writes, in a letter to Kahler in 1952, of the "vacuum" in which they lived there.[22]

There was no shortage of attempts to let the solemn commitment to the *new man* make a difference in the world. In 1948, Kahler, together with an array of intellectuals, addressed the public to introduce a draft constitution of a future world government. Robert Hutchins, for many years president of the University of Chicago, had assembled this "Committee to Frame a World Constitution," and he stated the alternatives. Here too, the tone is apocalyptic in mood: "one is world suicide; another is agreement among sovereign states to abstain from using the bomb."[23] Quite how seriously Kahler took this "Committee to Frame a World Constitution," is still expressed decades later in a letter to Else Jaffé in 1965:

> At the moment, I am involved in two utopias—feasible utopias inasmuch as they would both be desperately needed if it were to be still possible to save this world, that is, to keep it human. One concerns world organisation, and I am just on the way to California, where I will fly to consult with a group convened by the former president of the University of Chicago, Robert Hutchins. The other kind of Utopia is the old one I have been pursuing since *The Vocation of Science*, the integration of the sciences.[24]

The two were connected, for Kahler and the members of the committee saw themselves as a "nonpartisan advisory, pedagogical, and supervisory elite," which hoped to subject the world to the formative process of *Bildung*. "This elite," Kahler writes in an essay on "The Fate of Democracy," "would have the task, in our ever more technicized and functionalized world, to keep alive the spiritual as the genuinely human, the common sense of the experience of a people and of peoples." The mark of this humanistic elite is an "essential purity and superiority, which goes beyond [their] practical accomplishment, a sense and a care for the universal, for that which is common to the peoples and to humanity."[25] That is the barely altered metapolitics of the postponed *Bildung* of the human race, no less grandiose in its aims after years of exile. At the end of the 1950s he still speaks of "a total ethics," and he summarizes its maxims in the claim that "anything that leads to wholeness is good, anything that leads to split is bad."[26] But what exactly this *Bildung* of the *new man* was supposed to represent in terms of objects, books, canons or institutions was still uncertain, even after decades of exile.

It was not only in letters that Kahler himself repeatedly proclaimed the unity of his work, across the caesura of exile. He introduced his essay collection of 1952, *Responsibility of the Mind*, by confessing that the volume's essays form "a single unity, and that they are selected with this in mind. They all deal, directly or indirectly, with the great crisis of our time."[27] Clearly, Kahler had not formulated a fundamentally different self-understanding in exile. Rather, he was attempting, under the increasingly difficult conditions of exile, to uphold his high demands on himself, against all resistance, including his own attacks of depression and loneliness. For a short while, he may have had some expectations arising out of his relations with admirers at Ohio State University who were attempting to found an Institute of Humanism, where there might have been a place for him.[28] This would have been compatible with Kahler's own ambition to be more than just a college teacher. But the plans fell through. Perhaps such an opportunity would have provided him with a different situation in the United States, and with it a more accepting relationship with America and Modernity. As it was, he remained on the margin of American society, in his heart always closer to Germany. It was not necessary for him to learn the stylized outsider's role; that had already been taught him by the antibourgeois movements at the time of World War I.

To be sure, there was no lack of attempts at reorientation in exile. He tried repeatedly to move beyond the merely academic field of influence. With Albert Einstein, Kahler published a book in 1944 on *The Arabs in Palestine*, which supported older ideas of a binational state, already formulated by Kahler before 1933,[29] but which was timely and controversial in 1944 because of the imminent founding of the state of Israel. It goes without saying that Kahler took part, at an advanced age, in the anti-Vietnam Demonstrations. "It would be tolerable," Kahler writes to Thomas Mann in 1941,

> if the times were different. But as it is, there is nothing available against the deep sadness that overcomes one from the "wave of the future," against which we obviously can do so little. And although we know perfectly well that this *cannot* be the future of humankind, it may be ours, if something fundamental does not change here very soon, of which I regretfully see no sign. Things are very bad in this land—a most bitter "I-told-you-so." But enough of this.[30]

Kahler was relegated in exile to the role of undergraduate teacher. Apart from scattered lectures, the only chance of a wider impact came from the writing of books. *Man the Measure*, his principal American work, has had countless editions and translations since 1943. It is nothing less than an essay at a world history of humankind, a "total view." In the tradition of the "figuration view" [*Gestaltschau*] as practiced by the George Circle, Kahler attempts to reveal the secret order of meaning in history. No catastrophe, but it bears the *new man*. The course of historical events is thus contingent upon the "supra-temporal characterology,"[31] which is Kahler's aim. The unity of history tends to dissolve the subject matter of history. The demonstration through the philosophy of history of the essential *Bildung* of man paused over details only for the sake of illustration. In such moments, *Bildung* has objects, as with the development of the city. But they were functions of the superordinate figuration of history as a whole.

It is easy to show that even in this frequently reprinted book none of the extrapolated lines of historical development are new. Kahler had already described them in what was perhaps his best book, *German Character in the History of Europe*. This book rested on years of preliminary work between the wars and, because of Hitler's seizure of power, could be published first in Kahler's Zürich exile in 1937. Perhaps an earlier success of this book would have led Kahler back to work on history, and the expressionist pathos of *Bildung* might have been channeled into institutional and disciplinary paths. In exile, however, it was hardly possible for it to attract any readers. Its main thesis—that Germany, the unfinished nation, was the place where the world-historical contradictions would collide more directly than elsewhere—reappears scarcely weakened in the American adaptation of the book. The National Socialist seizure of power was bound to confirm him in this belief.

Before 1933, but also afterward, Kahler had tried to escape the aporia of extremist interwar philosophy of culture, and to put the case for a humanism turned toward life-orientated humanity. His work was supposed to be heir to the "old" *Wissenschaft* and to identify the "new" knowledge essential to the "kingdom of man." This ambition could not be achieved in this way because Kahler's ambition implied, as Arthur Salz had already observed in 1921, the postulation of a "teaching of wisdom" that demanded the "utopian" overturn of the entire tradition of science.[32] For this reason, Max Scheler catalogued Kahler's revolution in science among the movements "harmful to the presence of genuine philosophy and *Wissenschaft* in Germany," because it presumed "to attempt to make over, in its foundations as well as its methods, a cumulative achievement of two thousand years of occidental history—namely, rational, inductive or formally deductive disciplinary *Wissenschaft*, free of presuppositions of worldview."[33] Prior to 1933, this was utopian, and, all the more, hardly communicable in the United States, outside emigrant circles. Under the conditions of exile, which Kahler entered as a man of almost fifty years, and which would grant him nothing more than a supporting role in contemporary intellectual debate, exile supported a self-immunization of his own thinking rather than leading it out of the German provinces of meaning.

Was this development inevitable? In my opinion, yes and no. Yes, because Kahler remained more strongly devoted to the expressionist generation and its life-reforming uprising than he himself realised, even in exile. At fifty years of age, Kahler was simply too old to consider the world afresh. And exile only rendered the possibility of a reconciliation with bourgeois modernity more difficult. No, because it was not impossible that Kahler could have found an entry to the liberal traditions of civil society, especially in the United States. And no, because Kahler, for all his radicalism, had never spoken on behalf of hatred or ressentiment. Indicative of this is the distance with which Kahler remained disengaged from the fascinating George. He knew himself at one with Friedrich Gundolf about the "beautiful and simultaneously threatening feeling" of having "a closed view of the world,"[34] in Gundolf's formulation of 1915. No, moreover, because Kahler's work before exile and especially in his principal work on the "German character" builds crossings more than once to the modern social sciences and histories of mentality. In exile, these crossings were neglected. Kahler's apocalyptical philosophy of history was, accordingly, not inevitable. In the isolation of exile, however, it lay close to hand.

Along this way, the dream of a humanistic revolution of the old *Bildung* and its renewal could hardly have learned to identify the objects, methods and institutions of *Bildung*. Kahler was aware that precisely these items had to be identified, and even fought for it. At this point, however, his work remains silent, and it never learned in exile what was to constitute the *new man*. It is true that Kahler, if only for reasons of sheer survival, had written more in exile than in all the years before. But what was supposed to make up the new *Bildung*, what ways might have transmuted the endlessly growing knowledge into a key orientation, he could not say. Even in exile, *Bildung* remained an ethos of absolute dedication to the human. But what was to comprise this humanism, no one could say. In exile, Kahler dug himself far more deeply into the trenches of his almost mystic belief in the new empire of man than seems necessary, when viewed from outside. Germany alone and no place in exile remained the battleground where this world-historical moment of the *new man* was to achieve its breakthrough. With the years, this faith became a matter of necessity for him, in order to see for himself at least the watermark of the coming *Reich* shimmering through. That is where salvation lay, at least for Kahler himself.

Notes

1. Georg Simmel, "Der Begriff und die Tragödie der Kultur" [1911] (The concept and tragedy of culture), in Georg Simmel (ed.), *Das individuelle Gesetz* (Frankfurt: Suhrkamp, 1987), 116–147, 143.
2. Ibid., 94.
3. Barbara Beßlich, *Wege in den "Kulturkrieg." Zivilisationskritik in Deutschland 1890–1914* (Ways in the "Cultural War": Critique of Civilization in Germany 1890–1914) (Darmstadt: Wissenschaftliche Buchgesellschaft, 2000).
4. Friedrich Nietzsche, "Der Fall Wagner. Ein Musikanten-Problem" (The case of Wagner), in Friedrich Nietzsche (ed.), *Sämtliche Werke. Edition with a Critical Study in 15 Volumes* (München: Deutscher Taschenbuch Verlag, 1980), 6: 9–54, 39.
5. Cf. Kay Schiller, *Gelehrte Gegenwelten. Über humanistische Leitbilder im 20. Jahrhundert* (Learned counter-worlds. Humanistic model in the 20th century) (Frankfurt: Fischer, 2000). See also Karl Mannheim, "Heidelberg Letters: Soul and Culture in Germany," in David Kettler and Colin Loader (eds.), *Karl Mannheim's Sociology as Political Education* (New Brunswick and London: Transaction, 2002).
6. Erich Kahler, "Theater und Zeitgeist" (Theater and Spirit of the Times), *Jahrbuch für die Geistige Bewegung* 3 (1912): 92–115, 101.
7. Ibid., 110.
8. Gerhard Lauer, *Die verspätete Revolution. Erich von Kahler. Wissenschaftsgeschichte zwischen konservativer Revolution und Exil* (The Delayed Revolution. Erich Kahler. The History of Scientific Thought between Conservative Revolution and Exile) (Berlin, New York: de Gruyter, 1995), 73.

9. Erich Kahler, *Das Geschlecht Habsburg* (München: Der Neue Merkur, 1919), 10.
10. Ibid., 7.
11. Hugo von Hofmannsthal to Efraim Frisch, 28. Mai 1919, quoted by Max Kreutzberger, "Hofmannsthal und Efraim Frisch. Zwölf Briefe 1910–1927," *Hofmannsthal-Blätter* 5 (1970): 356–370, 361.
12. Publisher's prospectus, *Der Neue Merkur* 3 (1919).
13. Thomas Mann, "Erich von Kahler," *Deutsche Blätter* 3, 28 (1945): 18–20, 20.
14. Thomas Mann, *Tagebücher 1935–36* (September 28, 1935) (Frankfurt: Fischer, 1978), 180.
15. Erich Kahler, "Ordnung" (Ordering), in Erich Kahler (ed.), *Die Verantwortung des Geistes* (Responsibility of Spirit) (Frankfurt: Fischer, 1952), 19–23, 23.
16. Arnim Mohler, *Die Konservative Revolution in Deutschland 1918–1932. Ein Handbuch* (The Conservative Revolution in Germany 1918–1932: A Handbook) (Darmstadt: Wissenschaftliche Buchgesellschaft, 1989) (first edition 1950). Part of an investigation into Hofmannsthal's formula and here (quite problematically) in Mohler's construed genealogy.
17. Ernst Krieck, "Vom Sinn der Wissenschaft" (The Sense of Sciences), *Der neue Merkur* 5, 7 (1921): 510–514, 511.
18. Erich Kahler: *Der Beruf der Wissenschaft* (The Vocation of Sciences) (Berlin: Georg Bondi, 1920), 405.
19. Ibid., 21–22.
20. Ernst Troeltsch, "Die Revolution in der Wissenschaft" (The Revolution in Sciences), in Ernst Troeltsch (ed.), *Gesammelte Schriften*, vol. 4 (Tübingen: Mohr Siebeck, 1925), 653–677, 676.
21. Paul Tillich, "The Conquest of Theological Provincialism," in William Rex Crawford (ed.), *The Cultural Migration* (Philadelphia: University of Pennsylvania Press, 1953), 137–156.
22. Wolfgang Pauli to Erich Kahler, May 7, 1952, printed in Gerhard Lauer, *Die verspätete Revolution. Erich von Kahler. Wissenschaftsgeschichte zwischen konservativer Revolution und Exil* (The Delayed Revolution. Erich Kahler. The History of Scientific Thought between Conservative Revolution and Exile) (Berlin, New York: de Gruyter, 1995): 294.
23. Robert M. Hutchins, quoted in the *Saturday Review of Literature* 31, 14 (April 3, 1948): 6–8, 6.
24. Erich Kahler to Else Jaffé, November 25, 1965. (Deutsches Literaturarchiv Marbach).
25. Erich Kahler, "Das Schicksal der Demokratie" (The Fate of Democracy) [1948], in Ulrich Matz (ed.), *Grundprobleme der Demokratie* (Darmstadt: Wissenschaftliche Buchgemeinschaft, 1973): 35–65, 63–64.
26. Erich Kahler, *The Tower and the Abyss. An Inquiry into the Transformation of the Individual* (New Brunswick: Braziller, 1989) [First Edition 1957], 230–231.
27. Erich Kahler, *Die Verantwortung des Geistes. Collected Essays* (Frankfurt/M.: Fischer, 1952).
28. David Kettler, "The Symbolic Uses of Exile: Erich Kahler at Ohio State." Paper given at the Conference "Exile and Otherness. New Approaches to the Experience of the Nazi Refugees. Erich Kahler, *Israel unter den Völkern* (Israel among the Nations) (München: Delphin, 1933 [original destroyed]) (New print: Zürich: Humanitas, 1936)." April 29–May 2, 2004.
29. Erich Kahler, *Israel unter den Völkern* (Israel among the Nations) (München: Delphin, 1933 [original destroyed]) (New print: Zürich: Humanitas, 1936).
30. Erich Kahler to Thomas Mann, May 23, 1941, in *Thomas Mann—Erich Kahler. Briefwechsel im Exil*, in *Blätter der Thomas Mann Gesellschaft* 10 (1970): 18.
31. Erich Kahler, *Der deutsche Charakter in der Geschichte Europas* (The German Character in the History of Europe) (Zürich: Europa, 1937), 13.

32. Arthur Salz, *Für die Wissenschaften gegen die Gebildeten unter ihren Verächtern* (For sciences against the Learned under his Critiques) (München: Drei Masken, 1921), 5, 32, and 13.

33. Max Scheler, "Weltanschauungslehre, Soziologie und Weltanschauungssetzung," in Max Scheler, *Schriften zur Soziologie und Weltanschauungslehre*. vol. 1: *Moralia* (Leipzig: Neue Geist, 1923), 1–26, 9.

34. Friedrich Gundolf to Erich Kahler, December 21, 1915 (Deutsches Literaturarchiv Marbach).

CHAPTER SIX

AN EXILE'S CAREER FROM BUDAPEST
THROUGH WEIMAR TO CHICAGO: LÁSZLÓ
MOHOLY-NAGY

Anna Wessely

Exiles, Refugees, Cosmopolitan Nomads, and other Migrants

There is a characteristic two-stage pattern of exile in the case of Hungarians who fled their native country after the fall of the 1918–1919 revolutions, settled in Germany and then found themselves forced to flee again by Hitler's rise to power. I want to focus on the career of an artist, László Moholy-Nagy, and compare some of its turning-points with those in the life of a sociologist, Károly Mannheim. This comparison is intended to highlight the importance of international professional networks for migrants within the art world or academia, which account for some specific features that set apart the careers of scientists, scholars, and artists in exile from those of other emigrants.

I suggest that the all-too-broad concept of exiles, that sometimes apologetically accommodates "inner exiles" as well, should be broken down by distinguishing, among migrants, the categories of refugee, exile, immigrant, and cosmopolitan nomad.

Refugees are persons, forced by persecution (or disasters) to leave their native countries in search of security. Normally, refugees would prefer to return if conditions in their country of origin improve. This designation refers, then, to a temporary and transitory status that is followed either by the return to the native land or settlement in another country.

I would call *exiles* only those persons whose activities abroad are shaped by their intention to influence the political situation in their home country. Such people have either voluntarily chosen to leave their country or been forced by persecution to do so. They form governments in exile and negotiate with foreign government officials, reconstitute disrupted organizations or create new ones, publish newspapers, journals, pamphlets, or books not only in order to sustain the determination and spirit of solidarity among their fellow exiles but also for illegal circulation in the country they had left behind. Exiles tend to consider life in the host country as a finite, and hopefully short, phase in their careers. This implies the evasion of long-term commitments in the host country, little interest in its domestic politics if it has no impact on one's own country. Social ties to the "locals" are subordinated to political considerations and day-to-day life is defined by plans and preparations for the anticipated return.

An *immigrant* is, in contrast, a person who decides, for whatever reasons, to settle permanently in another country. Consequently, immigrant individuals and families are strongly motivated to improve their positions in the new country, to obtain citizenship, find recognition by the "locals" and to build and maintain various networks that facilitate their acculturation and social integration. Immigrant individuals leave behind their country of origin with no intention to return on a permanent basis.

Cosmopolitan nomads are not just a postmodern product, but well-known throughout Europe before the consolidation of nation states as missionaries, mercenaries, humanists or vagabonds. They relied on networks that spanned the whole continent. Somewhat similar networks were created again, from the late nineteenth century on, by the various revolutionary movements, most notably by the Comintern. Nomads easily communicate and adapt to new environments, but their long-term connections and commitments do not bind them to any particular country. They go where the action is, either as commanded by the people in the node of the network or simply wherever intellectual or political life offers support for their various beliefs and activities.

Refugees and exiles are defined by their country of origin, immigrants by their adopted country, while this issue of belonging seems irrelevant or, at best, of little import to cosmopolitan nomads. Moreover, refugees and exiles are pushed by the need to leave their native country, while immigrants and nomads are pulled by the opportunities offered by another country to pursue their ambitions, realize or continue the activities crucial to their life-plans. This typology is, of course, not absolute. Transitions among the ideal types outlined above are possible and very probable. A migrant person would often combine certain aspects of two or more types, giving prominence to one or the other aspect depending on changing circumstances as well as on the accumulated experiences of the years spent abroad. Nevertheless, it makes sense to separate these types analytically for they provide good approximations of the concepts underlying migrants' self-images and shaping the way they will be perceived by people with whom they associate in their new environment.

In order to explicate and justify this classification, I use the test case of the community of Hungarian refugees in Vienna in 1919. They fled there from the strongly anti-Semitic "white" terror wave of retaliations that killed or threatened all who had (or might have had) any part in the revolutions of 1918–1919.

Let me begin with *cosmopolitan nomads* who seem to have never experienced what is referred to as the modern intellectual's close tie to the nation, to its historical tradition and collective memory. Ilona Duczynska, the daughter of a Polish–Hungarian noble family in Austria, finished high school in Vienna, began her university studies in Zürich from where she was sent to Hungary in 1917 as a courier with the international socialist antiwar manifesto of Zimmerwald. She resumed her studies in Budapest and plunged herself into antiwar political activity until she was arrested and sentenced to many years of prison. The October 1918 revolution set her free, the

People's Commissariat of Foreign Affairs of the March 1919 revolution sent her on a mission to Switzerland. The next year she was already working in Moscow, aiding Karl Radek in the preparations for the Second World Congress of the Comintern. The next mission took her to Vienna where she met Karl Polányi in a hostel of Hungarian refugees. They married in 1923. Duczynska continued her university studies and worked for various left-wing Austrian journals. When Polányi left for England in 1933, his wife stayed on in Vienna as the editor of *Der Sprecher*, the clandestine journal of the famous workers defence militia, the *Schutzbund*. She joined Polányi in England (in 1936) and then in the United States (in 1941), from where the couple returned to England in 1943 since Duczynska, an expert on aeronautics, wanted to contribute to Britain's war effort by working for Royal Aircraft. In 1947 Polányi accepted a professorship at Columbia University in New York, but his wife, as a former communist, was refused an entry visa to the United States. She had to settle in Toronto and Polányi became a commuter. Together they edited an English-language anthology of Hungarian literature (1963). After her husband's death, she frequently visited Hungary where she soon found ways to support the developing leftist intellectual dissident scene. She died in Canada in 1978.

The point I want to make with this abbreviated biography is that the only time in her nomadic life that Ilona Duczynska regarded herself an exile was when, forbidden "for all time" from entering the United States, she had also discovered, after an extended visit to Hungary, that the country was less than inviting for revolutionaries of her cast. Then she finally decided to emigrate to Canada in 1950.[1]

Hungarian refugees who had played an active role in revolutionary politics, for example, Oscar Jászi or Georg Lukács were *exiles* who worked in Vienna to mobilize resources and foreign support against the counterrevolutionary government or to save the lives of imprisoned comrades. When all plans and efforts proved futile, Jászi decided to emigrate to the United States and become a full-time academic. Lukács joined the cosmopolitan communist network, living first in Germany and moving on to the Soviet Union. The years of World War II mobilized Jászi again, turning him into a part-time exile politician; Lukács, sent to work in Hungary in 1945 by the cosmopolitan Communist centre in Moscow, reluctantly found himself returned to his home country.

Political exile of a different sort is represented by the decision made by a common friend of their youth, Béla Bartók, who, persecuted in 1919 and afterward tolerated by the Hungarian authorities, decided in 1940 to move to the United States in protest against the pro-German policies of his native country.

That the Hungarian refugees in Vienna in the early 1920s represented a very diverse lot was evident to all participants and outside observers. One of them, Karl Mannheim, attempted in his "Letters from the Emigration" (published in Hungarian in 1924)[2] to survey their situation and break down the category of "exile" along substantive—political—criteria:

> The emigrant community is a theoretical concept to which nothing corresponds in organisational terms. And nothing can correspond to it, not even in Vienna where most of the emigrants live, for the various fractions of the emigration stand isolated and very far from one another.

He proceeds to distinguish mere refugees from exiles. The former are defined by their hopes to be permitted to return as repentant prodigal sons to Hungary; the latter, independently of whether they have been forced to escape or have left in disgust with the new regime, are for Mannheim the "genuine" emigrants—corresponding to what I have termed exiles. They are the intellectuals who do politics on the level of discourse. Their task, in Mannheim's words, their "national mission," is to represent and preserve the ideals and conscience of the people. The form of political action open to intellectual exiles is, accordingly, exhausted in refusing or granting legitimacy to a political system. This is expressed by their individual decisions to leave their native country and return only when that country deserves and learns to appreciate its intellectuals again:

> All emigrants were hurled out of the country by the revolution, but not all of them were revolutionaries. [. . .] Emigration is a hard lot, indeed, the school of chattering teeth and rumbling stomachs. Those who feel sorry [for their roles in the revolution] do not belong among us, that is, they are not genuine emigrants. Who is a real emigrant, then? Those who feel that the opposition between their worldviews and the regime is unbridgeable. Among them there are those who were not persecuted and could have stayed at home just as well, yet they could not remain precisely because of their sense of that opposition. Their emigration was voluntary but they have more in common with the forced emigrant revolutionaries than those who are now longing to return even if they have been forced to emigrate. The emigrants would or could return only if the present regime were liquidated [. . .] Until then the emigrant community should remain, for it has an important "national" mission. It has to save and keep alive the free soaring of the Hungarian mind and lend words to the conscience of the Hungarian people.

The fantasy of the intellectual exile about future recognition and praise is framed in a dramaturgy of apocalypse and second coming:

> And yet the liquidation of that regime is out of question. It is not even desirable. What has started to decay and rot must rot completely, its conservation would only impede the future healing of the Hungarian people. Let it rot to the bottom, to the end, and let the yearning for a universal, healthy renewal fill the whole of public opinion in the country.

This yearning will be the call to the exiles, "the true and good Hungarians who put a high value on the honor of Hungarians, who love our people more than those criminals and moral degenerates do," to return and secure "the reign of common sense and insight." Hungarian public opinion, rotten to the bottom, will miraculously heal and "adorn the emigrant community with the wreath of patriotism."

Wishful millenarian fantasies apart, the exiles had to leave the "school of chattering teeth and rumbling stomachs" and adapt to the host country in order to make a living. In short, if they did not feel attracted to the political network of cosmopolitan nomads, they had two options left. They could apply for return to Hungary in the second half of the twenties by which time the new regime had consolidated and allowed a limited pluralist space for intellectual practices to develop. Or they could remain abroad, accepting and adapting to the conditions of work and life the host countries offered them. Filled with ambition and the wish to prove themselves as the genuine intellectual representatives of the country they had left behind, the younger generation, eagerly sought out those centres and circles where they could learn, develop, and make a career. Most of them moved on to Germany.

Pressure to conform is always much stronger on the resident alien than the native citizen. Its constraints could be best avoided by attaching oneself to the most radical, critical and innovative faction within one's chosen profession in the new place. Such factions generally welcome new recruits who help to strengthen their positions in their struggle with the established conservatives in the field. The newcomers' unusual ideas and ways of thinking might also become a source of innovation for the whole group. For the emigrant, joining these progressive groups permits resistance to the pressures to conform as well as identification with and dedication to the goals of the domestic rebels. Thus the privileged intellectual fields of science, humanistic scholarship, and the arts open to immigrants a path to acculturation that not only tolerates but actually demands that they preserve their critical perspectives. This move allows for a universalist redefinition of their political stance on an intellectual level, while it miraculously transforms the social and political marginality of the exile into the self-consciously and heroically assumed marginality of the critic or innovator. The alternative strategy of joining the conservative faction within one's field is generally a less appealing, if not an altogether foreclosed, option for the immigrant, simply because conservatives tend to be strongly attached to the national traditions within their respective fields.

The Parallel Careers of Karl Mannheim and László Moholy-Nagy

In terms of the above suggested typology of migrants, I want to claim that Moholy-Nagy and Mannheim were exiles from Hungary and refugees from Germany to England. Their major professional development and recognition took place in Germany in the 1920s, which did not lead in their case to the immigrant's identification with the adopted country but rather to a universalized, cosmopolitan type of identification with the task of educating humankind for a more liveable future. These goals and the efforts to achieve them inevitably made both Mannheim and Moholy-Nagy targets of conservative or nationalist attacks in the course of the politically loaded Weimar debates on education. Many opponents decried the proposed reforms as alien interventions in German public life.

Born in 1893 and 1895, respectively, Mannheim and Moholy-Nagy were young and at the very beginning of their careers at the time of their first emigration. Mannheim had published a few articles while Moholy-Nagy had a small exhibition of his first paintings and drawings opened in the south-eastern town of Szeged. Both associated with radical circles and were accordingly recognized as promising talents by the cultural policy of the short-lived revolutionary regime that appointed Mannheim a university lecturer and bought three drawings by Moholy-Nagy.

Although practically of the same age and social—Jewish middle-class—background, "experiencing the same concrete historical problems" and having similar likes and dislikes, strictly speaking they were not of the same generation-unit in the sense

Mannheim used this concept.[3] The crucial formative experiences of their lives, although undergone at the same time, in the years 1917–1918, sharply differed. For Mannheim it was membership in an intellectually and emotionally highly charged, vibrating group of scholars and artists—the so-called Sunday Circle—for Moholy-Nagy, a law student drafted into the army, it was the shattering experience of the war.

Both Moholy-Nagy and Mannheim fled to Vienna after the fall of the republic of councils in Hungary and then went on to Germany in 1920. There they mastered the cultural difficulties and overcame the institutional obstacles that impeded the admittance of foreigners to prestigious positions. Investing their talents and evidently boundless creative energies in innovative cultural sectors ready to recognize their strengths, they became major figures in their respective fields in less than a decade. Thus they came to be seen as representatives of German cultural achievement by colleagues in other countries. This is how they were received when they embarked on the second stage of their emigration, this time to Britain. Mannheim stayed there while Moholy-Nagy soon followed an invitation to Chicago.

Occasional visits to Hungary in the late 1920s–early 1930s and contacts to Hungarian colleagues convinced Mannheim and Moholy-Nagy that the political conditions that had led to their exile had remained fundamentally unchanged and cultural or academic institutions were still unfavorable to the pursuit of their professional goals. Whether they would have ever seriously considered return to Hungary after the end of World War II, is impossible to say since both died relatively young: Moholy-Nagy in November 1946, Mannheim in January 1947.[4]

Art and Politics: Moholy-Nagy from Budapest to the Bauhaus

The biographical documents on Moholy-Nagy's youth in Hungary point to several important events and social relationships that formed his personality, influenced his political orientation, and shaped his conception of the social commitment of art as art. Born and raised in a village in the South of Hungary in 1895, he moved with his brothers to Budapest in 1913 where he began his studies of law. Here he met and made close friends with Iván Hevesy, a student of art history and editor of a literary journal, who was to become the first merciless critic of his poems and drawings (and, incidentally, also his godfather, when he decided to get baptized). In autumn 1914 he was drafted into the army, wounded, and appropriately decorated. During a long convalescence in a field hospital in Odessa, he found that drawing apparently became his form of psychological self-therapy.

Returning to Budapest in 1918, he spent there a little over a year before his emigration, but in this extremely intense period he divided his days between getting his doctorate in law, pursuing systematic studies of painting in the studio of Róbert Berény, attending lectures and discussions at the Galilei Circle (a progressive student organization within the Association of Free Thinkers, founded by Karl Polányi in 1908 with the aim to fight illiteracy and ignorance as well as to promote the education of the working classes), and making daily visits to the mortally ill Symbolist poet and radical journalist, Endre Ady. He exhibited water-colors at a show in the National Salon, made friends not only with Béla Balázs (organizer of the Sunday Circle meetings) but also with the most outstanding painters of his generation who involved him in the circle of Activist artists gathered around Lajos Kassák and his journal entitled *MA* (*Today*).[5] Kassák's journals in the 1910s belonged to a Europe-wide network of avant-garde art publications that mutually supported one another by exchanging materials. That Kassák tried to keep up this international orientation all through the war years is very important from the point of view of Moholy-Nagy's exile career for these contacts were to facilitate his admission to the Berlin art world.

The *MA* group reconstituted itself in Vienna: it organized lectures and perform-ances and relaunched the journal in 1920. The journal was forbidden in Hungary, but it circulated in the Hungarian communities of Romania, Yugoslavia, and Czechoslovakia. The first Viennese issue greeted the reader with a manifesto in Hungarian and German, addressed to the artists of the world (*An die Künstler aller Länder*). The experiences of their conflicts as Activist artists with the cultural policy makers of the Hungarian Republic of Councils (represented by the People's Commissar for Education, György Lukács) prompted the editors to distance themselves from the notion of a class-based proletarian art of the future.[6]

Moholy-Nagy felt uneasy among these people for several reasons. He was disgusted by the internal strife among the various suffocatingly close circles of Hungarian refugees who seemed to him to have lost sight of the political goals that had made them choose exile rather than submission to the governments appointed by Admiral Horthy. Perhaps he also feared the crushing effects of Kassák's domineering personality on his own development and preferred to coedit the journal from a safe distance as its Berlin correspondent. In a letter from Berlin to his friend, Iván Hevesy, also a former contributor to *MA*, he reported that the people around *MA*, although unable to move forward in their art, cared only for themselves with only a very few of them willing to work "for the cause itself," that is, for progressive politics.[7] Thus Moholy-Nagy was eager to move on to Berlin where he could learn, develop, and find contact with that international avant-garde network that was to sustain him and which he, in turn, helped to strengthen and extend in the next decade.

Having reached Berlin by earning his fare and food on the way as a sign painter, he soon found himself a studio and immersed himself in painting and in the art world of the city. Berlin after the war was dynamic and cosmopolitan, eager for every new idea. It welcomed Expressionists, whether steeped in mystical admiration of their own unfathomable souls or filled with dreams of a utopian future, both Gothic and Modern. It enjoyed pathos as well as sharp social criticism or Dadaist cynicism; it relished Malevich's deeply religious Suprematism no less than the rational engineering products of Soviet constructivism. The war-time contacts between the journal *MA* and Herwarth Walden's journal, *Der Sturm* as well as the presence of numerous Hungarian artists in Berlin allowed Moholy-Nagy to see everything, meet everybody, absorb and experiment with every novel idea and technique, participate in theoreti-cal debates, sign manifestos, etc.

Moholy-Nagy assimilated all that he had seen and learned into attempts to solve what he came to define as the problem of painting: How could one capture and activate light, this wonderful medium that by virtue of its power to create space through movement, reflection, and cast shadows could unite the constructivism of architecture, the flexibility of dance, and the visual harmony of balanced forms on the painted surface? How could one make light work in other media—metal surfaces, photographic plates or film? While his autobiographical "Abstract of an Artist" (written in 1944) elaborates in great detail how this time of fermentation led to discovering for himself the problem of painting and how these insights fed his confidence in his own observations, nothing suggests that as a painter he ever considered himself an exile.

The journal *De Stijl* published in 1921 a Dadaist manifesto (*Aufruf zur elementaren Kunst—An die Künstler der Welt!*), signed by Raoul Hausmann, Hans Arp, László Moholy-Nagy, and Ivan Puni), which demanded that artists produce innovative art, reject styles and fashions and throw off the fetters of utility and beauty. The manifesto in itself is not particularly informative or interesting, except for the fact that it shows the kind of cosmopolitan network that Moholy-Nagy, joined in Berlin. By 1922 he had become one of the few capable of shaping that everfluctuating network by marking its points of crystallization and defining its structure of relevance. His involvement with Hungarian radical political circles and art journals at the time was no remnant of the past or a mere provincial affair, but the activation of a node of the international avant-garde network with a particular purpose in view.

Interestingly, but understandably enough, the Hungarian articles Moholy-Nagy published or coauthored are much more political than anything he had written for a German or international audience. They represented the form of politics open to the exile who happens to be no citizen of any polity: a purely verbal radicalism that flows smoothly over into references to the vaguely outlined image of a desirable future society. The ideal of a communist society modern artists should work for was also offered as a vanishing point where the different perspectives of the various emigrant factions might peacefully merge. These writings also offered Moholy-Nagy an opportunity to elaborate his own conception of contemporary art and its social mission. The question of how art making could be shown to extend beyond the egotistic pleasure of the artist and could be justified as vitally important for the "masses" of needy people had occupied him already at the time when he decided to give up law for painting. At the time of the Republic of Councils in Hungary, he had entered the following thoughts into his diary:[8]

> During the war, but more strongly even now, I feel my responsibility toward society. My conscience asks incessantly: is it right to become a painter in times of social revolution? May I claim for myself the privilege of art when all men are needed to solve the problems of sheer survival?—Art and reality have had nothing in common during the last hundred years. The personal satisfaction of creating art has added nothing to the happiness of the masses.—I have had many talks with men and women on my long train trips. I have seen what is needed beyond food. I have finally learned to grasp what biological happiness is in its complete meaning. And I know that if I unfold my best talents in the way suited best to them—if I try to grasp the meaning of this, my life, sincerely and thoroughly—then I'm doing right in becoming a painter. It is my gift to project my vitality, my building power, through light, colour, form. I can give *life* as a painter.

Only three years later but worlds apart, the question he had to face concerned the social justification of abstract art to his communist friends. The open letter to the journal *Egység* (*Unity*) briefly surveys the history of modern painting from Impressionism to

Dadaism and then proceeds to repeat the earlier argument about the life-enhancing, vitality-increasing function of art:[9]

> And just as in our striving towards a new life we cannot be satisfied with the dogmas of existing societies, inasmuch as we are searching for the very laws of our humanity, so we had to start from scratch upon the work of determining the interrelationships of colours, forms and tensions (Suprematism, Constructivism). This is how we arrived in painting, for example, at the simplest and also the most complicated problems, the building up of a given surface by the ego, in which it is no longer so much the subjective experience that dominates but rather the objective demands of colour. This is the essence of the use of the much-ridiculed square. [. . .] The same applies in the case of spatial and material constructions. We seek the simplest solution that will provide the maximum possibility for treatment and for spatial tension, so that, on the one hand, man may learn to handle his materials, while on the other he may participate with his own tensions in his environment, thereby increasing its vitality.

The "Declaration," co-authored in 1923 with other Hungarian exiles—the artist László Péri (with whom he had a spectacularly successful exhibition in the Sturm Gallery the year before) and the critics Alfréd Kemény (later known as Durus) and Ernő Kállai—went a step further. It endeavored to distinguish the specificity of communist (Hungarian) constructivism from what the authors defined as the mechanized aestheticism of the Dutch De Stijl Group, on the one hand, and the technical naturalism of Russian constructivism, on the other, both of which they found to reveal "bourgeois traits." In a second step, they wished to demonstrate that Hungarian constructivism with its roots in communist ideology and its concentration on *relations* instead of *objects* was able to play a role "in the integrated process of social transformation." Incidentally, Moholy-Nagy succeeded in inserting into the text his idea of constructivist art as a politically legitimate alternative to work on "the collective architecture of the future, which will be the pivotal art form of communist society."[10] The "Declaration" was written with the intention to patch up misunderstandings and join efforts with the emigrant communist cultural journal, *Egység*. The offer was declined and the group broke up.

By this time, however, Moholy-Nagy had already joined the faculty of the Bauhaus in Weimar as its youngest professor. Invited by Gropius to lead the metal workshop, he also taught the preliminary course after Itten's resignation, edited the *Bauhaus Bücher*, and kept on working in traditional media and experimenting with new ones in order to find the appropriate forms for his transparent architecture of a relationally defined world.

Moholy-Nagy unconditionally identified with work and life in the Bauhaus. He discovered his true vocation in educating "whole persons" as opposed to the "segmental beings" that modern production and administration demand and specialized training delivers. The struggle of this unique educational institution[11] for survival culminated in a redefinition of its objectives. This led to the resignation of Gropius, followed by Moholy-Nagy, Marcel Breuer and Herbert Bayer. Moholy-Nagy's letter of resignation to the Bauhaus Masters' Council deplored the decision of the new school director to stabilize the position of the institution by strengthening its vocational character and promoting profitable production within the Bauhaus.

Moholy-Nagy saw in this the abandonment of the shared revolutionary goals and community spirit[12]

> The spirit of construction for which I and others gave all we had—and gave it gladly—has been replaced by a tendency toward application. My realm was the construction of school and man. [. . .] I am infinitely sad about this. It is a turn toward the negative—away from the original, the consciously willed, character of the Bauhaus.

A manuscript in his papers is even more clear on this point closest to his heart: "work within the Bauhaus was not purely of a laboratory nature; it was also pedagogical. The Bauhaus as a *new type of school* was of utmost importance."[13] Clearly, Moholy-Nagy's adamant support of the Bauhaus educational philosophy was not just the defence of a democratic form of education,[14] but as such also a political statement—less dramatic, perhaps, than the verbal radicalism of his earlier Hungarian-language publications, nevertheless, the optimal medium of politics open to a resident alien.

Karl Mannheim in the German Academic World

The Hungarian emigrants in Germany depended at first on the charity of friends, friends' friends, or benevolent strangers. Mannheim was supported by academics. He lived at first with Vilmos Szilasi in Freiburg, then he stayed with Emil Lederer in Heidelberg until he could find a job at a Darmstadt high school.

More important than sheer survival was, however, the mobilization and extension of personal networks that could secure the newcomer's admittance to the circles distributing social recognition in one's chosen profession. Through recommendations by Lederer and Lukács, Mannheim was invited to Marianne Weber's salon that he was careful enough regularly to attend. Social acceptance by Max Weber's widow amounted to a certificate of one's intellectual existence and eventual value as a sociologist. Norbert Elias's recollections of his student years in Heidelberg aptly summarize the crucial role of Marianne Weber's salon in making and unmaking careers as well as Mannheim's was full awareness of its importance in improving his own chances for a university position.[15]

Mannheim succeeded not only in getting integrated in German academic life, but he also rose to prominence within the sociological profession. His *Habilitation* was accepted on condition that he apply for German citizenship, which he was granted in 1927. The next year he was invited already as a keynote speaker to the annual meeting of the German Sociological Association and in 1930 he was offered the chair of sociology at the University of Frankfurt.

The fruits of the first decade spent in Germany were published in book form by Mannheim as well as Moholy-Nagy in the year 1929. *Ideologie und Utopie* and *von material zu architektur* represent not only the zenith of their careers in Germany, they are also among the best cultural achievements of Weimar Germany. While for Mannheim the following years until his dismissal from Frankfurt were filled with lecturing and elaborating the implications of his study of Weber's work for cultural sociology, Moholy-Nagy faced the most critical period of his life. In the following sections I want to show that, although Moholy-Nagy was reluctant to leave Germany, by 1934 it had become a matter of his survival as an artist.

The Artist Alone

In 1929 Moholy-Nagy left the Bauhaus in Dessau and moved to Berlin. This seems to have been a much more difficult and painful decision than leaving Hungary was ten years before. In the terms introduced at the beginning of this paper, I would say that now Moholy-Nagy became an exile, indeed, namely an exile from the Bauhaus. He remained attached to the Bauhaus idea all his life. This strong attachment turned him later into a cosmopolitan nomad eager to move wherever there was a chance to transform the Bauhaus ideal of art and education into actual practice again.

In Berlin, he opened a design studio and worked as designer of exhibitions and stage sets (for Piscator's theatre, the state opera and Kroll's opera house) as well as a photographer, film maker, typographer, and commercial designer. The studio prospered, but the man and the artist were heading for a crisis.

The clearest symptom of the crisis was his inability or unwillingness to continue painting in 1928–1930. His monographer, Krisztina Passuth, suggests that Moholy-Nagy's career as a painter had reached its peak by 1926. Unable to proceed on the path of systematic innovations, he turned, therefore, from painting to essay writing, stage design, photography and film. When he resumed painting in 1930, he was no longer an avant-garde artist but merely one of the many modernist painters.[16]

The inability[17] to paint for two years was, I think, Moholy-Nagy's form of mourning over the failure of the Bauhaus project and his personal loss of the sustaining world of Bauhaus activities, art and life in which his systematic experiments with vision and his innovations to overcome the illusionist rectangular flat surface of the traditional canvas directly made sense and assumed significance in a shared way of thinking. He had to locate a new purpose, a socially meaningful goal for painting to achieve, before he could return to it with the enthusiastic dedication of old. This moment arrived only after his emigration from Germany, when he had an opportunity to survey the whole of his painterly production in a spectacular exhibition of his works in the Stedelijk Museum of Amsterdam in 1934. There he realized that his approach to mobilizing the spectators' optical consciousness had not been radical enough; his works were mere cabinet paintings for sensitive, contemplative art lovers. What was needed, instead, was the adaptation of painterly innovation to the changed visual environment of modern life, that is, stronger effects to shock viewers into attention and emotional elaboration of what was taking place in the pictorial space in front of their eyes.[18]

Moholy-Nagy himself offered yet another explanation for abandoning and then returning to painting. In a letter to the editor of *Telehor*, the architect František Kalivoda, he felt compelled to expound his reasons for returning to painting of which he used to say that its means were "insufficient for the new requirements of art at a time when new technical media were still waiting to be explored." He listed several reasons, among them the lack of sponsors for experiments in the new media. His main targets were, however, journalism and public opinion with their sensationalism and cheap interpretations that blunted the sensibility of the public, transforming its "passionate desire for participation" into "the average newspaper reader's 'interest.' " The source of all these problems was, however,[19]

> to be found in the rapid spread of industrialism and the fact that it has been forced into the wrong channels by our capitalist production. [. . .]Every attempt to create a planned

economy, to reconstruct our uncontrolled, industrial world on a socialist basis, even every attempt at enlightenment, necessarily encounters the conscious or instinctive resistance of the ruling caste of society. And for the same reason every creative achievement, every work of art prognosticating a new social order and striving to restore the balance between human existence and industrial technique, is categorically condemned. The relatively meagre results of new experiments in art are due to the social system that rules our existence, the hidden ramifications of which actually extend to those circles where one would expect to find hostility to it.

In these circumstances, artists had to find alternative, almost devious ways to reach the sensibility of their prospective audience. Easel painting could be one of them by choosing to present artistic achievements in "small, carefully selected instalments."

Just how deep the crisis following the failure of the Bauhaus experiment and his own trauma of separation from it was can be best seen in a minor line of Moholy-Nagy's production from the years 1929 to 1933. In commercial art, where he could not expect his efforts to contribute to a larger-scale social project, he seems to have completely lost the ground under his feet. By proposing to analyse the stages of this process, I want to suggest that the lack of the sense of belonging to a socially important movement in the arts, together with the constraints imposed on a commercial artist by the political and trade interests of his patrons, had such a crippling effect on Moholy-Nagy's work as an artist that emigration became for him a personal and artistic necessity, independently of whether his life was in actual danger or not.

The Independent Commercial Designer in an Increasingly Political Media Market

On his return to Berlin in 1929, Moholy-Nagy was no longer considered by prospective patrons as one of the immigrant crowd, but rather as a famed professor of the most progressive school of design in Germany, the artist mainly responsible for the "Bauhaus look" in publishing and advertising. He had demonstrated and explicated what he considered the basic principles of up-to-date publicity a few years earlier, in the 1926 Bauhaus issue of the printers' journal *Offset—Buch und Werbekunst*. Here he published two short articles. The first confronted the goals of contemporary typography with aesthetically less satisfying modern practices; the second discussed the commercial use of the photographic montage in posters, advertisements and journals. The argument tried to bridge the gap that divided his abstract conception of art photography that was to be based on the one hand, solely on the effects of light contrasts, excluding all literary or imitative elements and on the other, on the advertisers' demand for an emphatic, eye-catching presentation of recognizable commercial products. Moholy-Nagy suggested that his montages, called photoplastics, were able to create an apparently organic and transcendent reality as well as evoke the desired associations in the beholder. Photoplastics were to offer visually witty combinations of photographical fragments without inserting any manually executed, drawn or painted figurative elements in the composition.[20]

When the publisher of the fading fashion magazine *Frauenmode* decided to replace it with a new journal targeting a self-consciously modern, urban upper-middle-class female readership, it was a reasonable decision on his part to entrust the design of the new publication to the best typographers of the Bauhaus school. He invited László Moholy-Nagy to design the layout and typography of the journal to be called *die neue linie*, and he appointed as art director Herbert Bayer, a former Bauhaus student and teacher, who had been by this time already in the employ of the British advertising agency, Dorland. The covers of the monthly issues of *die neue linie* were alternately designed by several artists, most frequently by Bayer himself, the Berlin arts professor, Otto Arpke, and László Moholy-Nagy.[21]

A closer look at the Moholy-Nagy designed front cover of the first issue in September 1929 (see figure 6.1), may reveal what struck the contemporary eye as "absolutely modern." The composition is full of multivalent, ambiguous elements. It

Figure 6.1   László Moholy-Nagy: Front cover of the September 1929 issue of *die neue linie*

shows a woman, an emblem of upper middle-class elegance, as both subject and object of the gaze, but inscrutably aloof in both aspects. Clothed in a carefully cut, fur-collared top coat, wearing a hat and high-heeled shoes, she turns her back to the viewer in order to inspect a huge wall in front of her. The wall is covered by rectangular photographs that are separated from one another by a white grid. The photos are all of snow-covered mountains with waterfalls and patches of rock. At some points, the grid seems to stand before a larger photo of which it merely conceals certain parts and thus separates its otherwise continuous elements; at other points it seems to belong to the plane of the obliquely set up white wall, evenly decorated with landscape photos of equal size. To the right, there is an orange-colored vertical plane, approximately perpendicular to the white one. It recedes into the depth of the pictorial space. Although this perspectival illusion is strengthened by the signature *moholy* on its lower edge, it is negated by the lettering in its topmost section. These white letters on the orange ground are involved in a strange quarrel: the smaller letters clearly indicate the date of the publication, September 1929, while the larger ones above them (the last three letters of the title *die neue linie*) say "*nie*," that is, "never" in German. The white inscription of the title itself defies the spatial positions of both the white and the orange planes: it hovers before them, even if this cannot be always distinctly seen when it is blotted out by the white grid. The woman stands in the corner formed by the white and orange planes. She is offered an enticing view that is too near to her to be gathered in a single glance, yet it does not overwhelm her, since the geometrical grid robs the image of the mountain landscape of its sublime grandeur. The overall effect is a curious combination of physical entanglement and emotional detachment, which reiterates the juxtaposition of painterly and photographic elements.

Opening the journal, the reader is addressed by the editors as a female member of the higher social strata, as *gnädige Frau*. She is urged to recognize herself in the ideal held up by the editors: A "real lady" must understand that "the good old times" are gone forever, nonetheless she should remain faithful to her inherited notions of exclusive elegance and reject the foreign and false ideals of Hollywood. The journal will point her toward a "third way" in fashion. This is "the new line" (*die neue linie*) that avoids both the self-styled clothes of the "roaring twenties" and the fake luxury of the "American look." It breaks with the pragmatic functionalism of the postwar years in order to rediscover femininity and pleasure in fine fabrics and ornaments, in distinction and elegance. Fashion in clothes will be in focus, but the journal also devotes separate columns to travel and outdoor sports (hunting, skiing, golf), entertainment (theatre, books, the latest gramophone records), and to what the editors term "the woman's sphere" (i.e., interior decoration, housekeeping, gardening, cooking, health, and beauty care).

While the fashion designs in the journal emphasize middle-class common sense, the sections on modern housing (*Neues Bauen*) and interior decoration present the journal as an advocate of modern ideas. Thus the first issue shows two one-family houses by Gropius and Poelzig each, discusses Mies van der Rohe's installation of the German textiles exhibition in Barcelona; the second describes the ideal home of the young professional single woman, and so on.

The Bauhaus attitude toward commercial design had been ambivalent from the very start. While the school was founded with the objective of revitalising handicraft and reconnecting its products within a unifying architectural environment, the exigencies of financial survival and political legitimacy soon forced it to embrace the latest technological principles of serial industrial production. The early slogan about the new union of art and handicraft was replaced by the demand for "a new union of art and technology."[22] While the painter masters in the school decried this turn as the victory of a schematic mentality (Klee called it *Schablonengeistigkeit*), Moholy-Nagy was among the younger masters eager to explore this new territory and confront its challenges. By the late 1920s the Bauhaus had, in fact, succeeded to produce marketable and fashionable prototypes for the industry in furniture, lamps, wallpapers, and fabrics as well as to become one of the leading lights in the field of printed publicity materials.

Fashion in clothes lay beyond the pale of the industrial design reform championed by the Bauhaus. While systematically and rationally reshaping architecture, interior design and the objects of everyday use, that is, the whole built environment, and thereby creating new spaces for the lives of modern men and women, the Bauhaus masters avoided direct interference with these lives as well as the modelling or clothing of the human body itself. Most Bauhaus artists preferred to leave it to the users to complete and individualize the objects designed for them. Moholy-Nagy's statement on this subject recalls the grand ideas, expressed already early in his career, which insisted on a social transformative mission of art:[23]

> Man must first understand that, within the productive community, he must crystallise a life community of people and within that life community, his own life. Only in this way can he grasp the true meaning of technical progress. The object of our genuine efforts should not be the *form* as such or the overwhelming technological process itself but rather a healthy *life plan* for man.

The journal *die neue linie* underwent marked change in the course of the first five years of its existence from the years of the Depression to Hitler's ascendance to power. The drift to the right in the editors'—and their potential readers'—political attitudes was at first hardly perceptible but then manifest in the articles.

Changes in editorial policy began already in the first year. The report of a correspondent in the February 1930 issue, which praises the colorful and sensuous ornamental riches of Viennese homes in contrast to the minimalist and technical aesthetics of the radical apostles of the Bauhaus style, indicates only a shift in taste. An expressly political reorientation is signalled, however, by the March issue devoted to Italy, the country of victorious fascism. Italian society is represented by aristocratic ladies and Italian interior design is illustrated by a report on the Villa Torlonia where Mussolini "lives with his dogs so near to his heart."[24]

Another, no less ominous political move of the editors was the introduction of a new column, beginning with September 1930: Under the heading "Why does *die neue linie* show . . . ?" the photos of artists, writers, filmmakers, and the like are shown up as worthy of praise or targets of suspicion and critique. The list of the condemned artists begins with Bert Brecht and continues with Max Reinhardt, Ernst Lubitsch (in 1931), Ödön von Horváth, Erich Kästner (in 1932), Arnold Zweig, F. Molnár, and Thomas Mann (in 1933–1934).

The 1933 volume hastened to greet the rebirth of German theatre "from the spirit of the national community" in response to the revolutionary dissolution of classes (June 1933); it upheld as the model of a modern home the house of the new Prussian

Prime Minister, Göring, redecorated by the portrait-painter, Dr. Elisabeth Keimer, who was proud to count, beside Göring, also Mussolini and the Pope among her sitters (August 1933); it urged its readers to buy only German products and regard it a patriotic duty to have their dresses made of expensive, fine clothes and thus contribute to the country's prosperity ("Reflections on Fashion," October 1933); it published a long essay by Peter Behrens that urged German architects to follow the example of Italian fascism in creating a national architectural style capable of expressing the new political ideals (November 1933); it invited a group of critics to decide on the best book of the year that happened to be, of course, "With Hitler to Power" (*Mit Hitler an die Macht*) by a certain Otto Dietrich.

Nevertheless, what the new political situation demanded was not so clear to everyone as it seems in retrospect. Art critics, for example, fought against "patriotic kitsch" and, not knowing what would later be defined as *entartete Kunst*, described works by Heckel, Kokoschka, Dix or Feininger in the July 1933 issue as "an almost national-revolutionary reappropriation of the essential traits of German existence."

The right drift of the journal also affected its visual appearance. The self-consciously modernist typography was gradually marginalized by a growing proportion of traditional typefaces, the cover designs became increasingly schematic and banal in style. The stylistic changes and confusion manifest in the ten front covers Moholy-Nagy made altogether for *die neue linie*[25] provide the best evidence for my thesis that emigration from Germany had become for him a question of survival as an artist.

The woman as object and subject of the gaze, as we have seen her on the front cover of the first issue, reappears several times (February and June 1930, July 1932). She is turned, however, into pure spectacle in the December 1931 (see figure 6.4) and the January 1933 covers. An even sharper contrast separates the earlier images— where the woman appears as a sovereign user of modern appliances or, at least, as their self-conscious adjunct (see figures 6.2 and 6.3)—from those last title pages for *die neue linie* where she merges into nature, loses her individual features and becomes a symbol for some vague idea of natural decay and rebirth (October 1932, figure 6.5). Finally, the cover design of the May 1933 issue (figure 6.6) shows how Moholy-Nagy slipped into a completely trivial pictorial idiom, losing that innovative edge and experimenting spirit that made his previous designs as well as his feverish activity as a painter and a teacher in the years of his second emigration (to Britain and, then, to the United States) so exhilarating.

The female figure in latter two images is reduced to a simple outline drawing in white with no features and expressive qualities of its own. In October 1932 (figure 6.5), it has become a mere natural sign that accompanies the change of seasons. Autumn is evoked by the prints of brownish dirty fallen leaves that a wind-blown dancing female form is unable to bring to life. This sense of helplessness or uselessness is even stronger in Moholy-Nagy's last work for *die neue linie* (figure 6.6). It is of such an amateurish quality that no monographer of this artist will reproduce or mention it at all. The image is to represent spring and it does this with the most worn clichés one can think of. The master of photoplastics produced here, on the front cover of a fashion magazine(!), an aquarelle of a green meadow with children dancing a roundelay and a young woman leaning onto a tree to the left. Her nondescript figure is clad in a nondescript suit. She wears a headscarf and with both hands holds a basket. The picture looks like an illustration transplanted from an elementary school book produced

Figure 6.2   László Moholy-Nagy: Front cover of the December 1930 issue of *die neue linie*

by closely following the thematic restrictions and stylistic prescriptions of what began to be called at this time, thousands of miles away, socialist realism.[26]

Internationalization and Emigration: Berlin, Amsterdam, London, Chicago

Moholy-Nagy tried to counterbalance the loss of the sustaining environment of the Bauhaus by intensifying his international contacts. These included contacts to his native country. He was regularly contributing to two Hungarian socialist cultural journals: *Munka* (Labour), edited by Lajos Kassák in Budapest and *Korunk* (*Our Era*), edited by Gábor Gaál in the Transylvanian city of Kolozsvár/Cluj. In 1930 he

Figure 6.3 László Moholy-Nagy: Front cover of the May 1931 issue of *die neue linie*

was invited to give a lecture, illustrated with slides, on the new view of the world. The event took place in a private gallery in Budapest. In its wake, the art historian Máriusz Rabinovszky published a study on Moholy-Nagy's photography in the leading Hungarian art journal (*Magyar Művészet*).

Moholy-Nagy also became the photography editor and regular contributor of the review *i 10. Internationale Revue* in Amsterdam and a member of the international movement of abstract artists, *Abstraction-Création*, founded by van Doesburg in Paris in 1931. In the same year, the Delphic Studios gallery in New York organized his first one-man show of photography in North America,[27] following the publication of the English translation of his book *von material zu architektur* (1925) with the title *The New Vision* (New York, 1930). These contacts could not mitigate his sense of

Figure 6.4 László Moholy-Nagy: Front cover of the December 1931 issue of *die neue linie*

isolation, moreover, his foreign colleagues reproached him for his attachment to Germany. He had to apologise for participating in the exhibition of the German *Werkbund*, organized by Walter Gropius within the Paris exhibition of the *Société des Artistes Decorateurs*, boycotted by French modernists. Former artist friends attacked him already in 1932 for not leaving Germany, but he was reluctant to move. He was still in Berlin in January 1934, helping persecuted people, sheltering them in his studio.[28] Soon it became clear that, even if he did not feel to be in danger himself (prior to his move to Chicago he repeatedly returned to Germany[29]), it was impossible for him, a modernist artist, to stay and work in Germany. His letters from Berlin

Figure 6.5 László Moholy-Nagy: Front cover of the October 1932 issue of *die neue linie*

to friends in England (to Herbert Read in January, to Barbara Hepworth and Ben Nicholson in April 1934) complain of a "frightening loneliness":[30]

> The situation of the arts around us is devastating and sterile. One vegetates in total isolation, persuaded by newspaper propaganda that there is no longer any place for any other form of expression than the emptiest phraseology. No wonder that one can barely bring oneself to assert one's influence, even in the smallest circle. One is forced into an insane solipsism.[31]

This spell of solipsism was broken by the decision to emigrate. He moved first to the Netherlands (1934), then to England (1935) and, finally, to the United States

Figure 6.6   László Moholy-Nagy: Front cover of the May 1933 issue of *die neue linie*

(1937). In retrospect, it looks as if he had been checking out several countries and experimenting with various forms of exile.

In the Netherlands, where he spent about fifteen months, he was employed full-time by a large printing house in Amsterdam and a silk-manufacturer in Utrecht, leaving hardly any time for his art work. Although he had a major retrospective show of his paintings and was highly appreciated as an artist, there was no space for the collective experience of creation and development that teaching at the Bauhaus had yielded.

In 1936, he returned to Hungary as a tourist. He found both the politics and the art world too depressing to seriously consider resettling in his native country.

He lived then in London for two years. Here he immediately found both large-scale work as a designer and window-dresser, which provided the means and facilities to continue his experiments with new materials and techniques. He also found his way to the art world, mainly through the artists of *The Circle* (Evans, Gabo, Hepworth, Moore) who arranged Moholy-Nagy's retrospective exhibition of his work at the London Gallery in 1936. The next year was filled more or less with shooting documentaries and the special effects for Alexander Korda's film, *Things to Come*, with the publication of three photography albums (*The Street Markets of London, An Oxford University Chest, Eton Portrait*), and with abortive plans to recreate the Bauhaus in England. So much "hectic, disjointed work" made him restless and dissatisfied. He missed the exhilaration of teaching and planned, with Gropius and Bayer, to open a Bauhaus in England. Lack of funds frustrated all such plans. The invitation by the Association of Arts and Industries to transplant the Bauhaus experience into the United States by establishing a school of industrial design in Chicago came as the perfect solution to his problem. It promised work again that "made sense" and could, if successful, "make history."[32]

Arriving in Chicago in 1937, Moholy-Nagy was neither a refugee, nor an exile but a cosmopolitan nomad attached to the international avant-garde movement, who hoped to find a home not so much for himself and his family as rather for his vision of a humanistic education in design as research into new forms of individual and social existence.[33]

The industrialists who funded the "New Bauhaus: American School of Design" had certainly different expectations. They wanted a school to teach future design consultants how to enhance the quality and competitive advantage of American industrial products.[34] Inevitably, this divergence in views concerning the objectives of the school led to a series of conflicts. These are generally interpreted in the literature as manifestations of the clash between American commercial pragmatism and European artistic idealism, although they merely reiterated with some significant modifications, due to the different traditions in the organization of production and marketing, the conflicts the Bauhaus had already experienced in its relations to the *Werkbund* and German industrial interests or Moholy-Nagy himself in his conflict with Hannes Meyer, successor to Gropius as director of the Bauhaus. Victor Margolin suggests, for example, that the Association of Arts and Industries could have better promoted its own interests if it had chosen to take over the successful curricula of the Carnegie Institute of Technology or the Pratt Institute, which were training design consultants who shaped their work in response to industrial clients and proved themselves capable of improving product sales. Unfortunately, the Association decided to transplant the Bauhaus model to America, because its board members were deluded by the cultural prestige and myth of the Bauhaus as an institution where artists invented new forms for industrial production. They realized only later that this phrase actually announced the designer's claim to lead production, to produce culturally more advanced prototypes for "progressive" manufacturers to buy and have adapted to mass production.[35]

The abrupt decision of the Association of Art and Industries to close down the New Bauhaus after a year of impressive and highly publicized teaching caught Moholy-Nagy unawares. Nevertheless, the experiences in Germany as well as in

America gave him the self-confidence that he would be able to negotiate this situation and bridge the gap between his conception of design education and sponsors' demands. His idea was that the school should function as a site for the education of whole persons who would learn to recognize and develop all their abilities and become capable of envisioning new forms of social life as well as inventing the objects appropriate to that life. The patented inventions that would arise as a by-product of work at the school would demonstrate to sponsors its usefulness for the industry.

Convinced of the rightness of his educational program, he tried to save the school or attach it to some other educational institution. When this attempt failed, he decided to found his own School of Design with the support of corporate sponsors. It opened February 1939. Part of the faculty he had recruited for the New Bauhaus (from emigrated members of his European network and philosophers from the University of Chicago) agreed to work on without a salary until the financial position of the school got stabilized. Moholy-Nagy supported himself by working as a design consultant for several large firms, among them the Parker Pen Company.

Hardly had the school consolidated its position when the war broke out and the conditions of the school's operation drastically changed. With reduced numbers of professors and students, involvement in defence preparations and meagre funds, Moholy-Nagy was able to maintain the school only by making more and more cuts and concessions.

The adaptation to wartime conditions in America entailed that the School of Design had to cut back its original program of art education and open toward the "commercial arts" of product styling and publicity. Its management and administration had to be brought in line with the more formal, hierarchically organized structures of American institutions of higher education. In 1944 it was renamed the Institute of Design.

In spite of the setbacks, Moholy-Nagy did not give up. He kept painting, teaching, experimenting with new synthetic materials and technologies. And just as he had been able to sum up the principles and experiences of his Bauhaus period in *The New Vision*, his last book, *Vision in Motion* (finished by 1945, but published already posthumously in 1947) set a monument to the Institute of Design by presenting the work of its teachers and students integrated into the history of the European avant-garde.

In September 1937, when Moholy-Nagy arrived to create a New Bauhaus in the United States, *The New York Times* introduced him in an article, entitled "America Imports Genius," as one of those famous German refugees whom America "helps to live and function at their favorite pursuits." In turn, the country expected them to become a ferment to its cultural life. The journalist added a cautionary note:[36]

> The hospitality that America extends to these men should not be merely physical, but spiritual. We should not be in too great haste to "Americanize" them—in the sense of attempting to indoctrinate them with all the beliefs we already hold. To make the most of their presence here we must think not only of what we have to tell them but of what they have to tell us.

The Institute of Design did not revolutionize American product design; the chances for setting up an international "parliament of social design" are as utopian today as at the time when Moholy-Nagy envisioned its establishment in the proposal with which he concluded *Vision in Motion*. Nevertheless, the exhibitions of the work of his Chicago students and of his students' students in the past three decades demonstrate that at least some Americans did listen to what László Moholy-Nagy was telling them. Moholy-Nagy died of leukaemia in November 1946. Mannheim, after a no less busy decade spent in teaching, writing, editing, and organising in Britain, survived him by a few months. Both of them were eager not simply to prove themselves but to offer what they thought to be invaluable service for the sake of postwar reconstruction: education.

Notes

1. Kenneth McRobbie, "Ilona Duczynska: Sovereign Revolutionary," in K. McRobbie and Kari Polanyi- Levitt (eds.), *Karl Polanyi in Vienna. The Contemporary Significance of the Great Transformation* (Montréal, New York, London: Black Rose Books, 2000), 255–264.
2. Karl Mannheim, "Levelek az emigrációból," *Diogenes* (Vienna), 1924, nos. 1–2, reprinted in Éva Gábor (ed.), *Mannheim Károly levelezése 1911–1946* (Budapest: Argumentum, 1996), 244–248. Quotations from this text below are in my translation. A German translation can be found in: Reinhart Laube, *Karl Mannheim und die Krise des Historismus* (Göttingen: Vandenhoek & Ruprecht, 2004), 582–593.
3. Karl Mannheim, "The Problem of Generations," in *Essays on the Sociology of Culture* (London: Routledge and Kegan Paul, 1956), 304.
4. In the autumn of 1946, both Mannheim and Karl Polányi were offered professorships by the University of Budapest.
5. Published in Budapest (January 1917–July 1919) and relaunched in Vienna (January 1920–1925).
6. Quoted by Júlia Szabó, *A magyar aktivizmus története* (Budapest: Akadémiai, 1971), 62.
7. Ibid., 388.
8. Quoted in Passuth, *Moholy-Nagy*, 386–387.
9. Ibid.
10. Kállai—Kemény—Moholy-Nagy—Péri, "Nyilatkozat" (Declaration), *Egység* 4 (1923), quoted from the English translation in Passuth, *Moholy-Nagy*, 288–289.
11. See the excellent monograph by Éva Forgács, *The Bauhaus Idea and Bauhaus Politics* (Budapest, London, New York: Central European University Press, 1991).
12. Published in Moholy-Nagy, *Experiment in Totality*, 46–47.
13. László Moholy-Nagy, "unterhaltung zwischen einem wohlgesinnten kritiker und einem vertreter des bauhauses weimar-dessau," quoted from the English translation in Passuth, *Moholy-Nagy*, 400–401.
14. Interestingly, Karl Mannheim also invoked the Bauhaus tradition in his best known discussion of sociology as a decisive medium of education in a democratic age.
15. Norbert Elias, "Notizen zum Lebenslauf," in Peter Gleichmann et al. (eds.), *Macht und Zivilisation. Materialien zu Norbert Elias' Zivilisationstheorie 2* (Frankfurt: Suhrkamp, 1984), 9–82.
16. Passuth, *Moholy-Nagy*, 52–53, 62.
17. Inability rather than refusal, for in a letter from Berlin to Will Grohmann, dated September 26, 1928, he expressly speaks about his "renewed interest in becoming a painter again," quoted by Gudrun Wessing, "Als Lichtvisionen von Schatten überholt wurden. Zur Situation der Bauhäusler um 1933 und Moholys Neubeginn nach den Jahren der Emigration" (hereafter cited as *Als Lichtvisionen von Schatten überholt wurden.*), in Georg Jäger and Gudrun Wessing (eds.), *Über Moholy-Nagy. Ergebnisse aus dem*

Internationalen László Moholy-Nagy Symposium, Bielefeld 1995. (Bielefeld: Kerber Verlag, 1997), 55 (hereafter cited as *Über Moholy-Nagy*.)

18. This is suggested by Sybil Moholy-Nagy in her book *Moholy-Nagy: Experiment in Totality*, quoted above.
19. "Letter to Kalivoda," *Telehor* (1936), English translation in Passuth, *Moholy-Nagy*, 332–335.
20. László Moholy-Nagy, "Zeitgemäße Typografie (Ziele, Praxis, Kritik)" and "Fotoplastische Reklame," *Offset- Buch- und Werbekunst* 7 (1926): 375–394.
21. All reproductions of the covers of the monthly issues of *die neue linie* are credited to the Bildarchiv Preussischer kulturbesitz/Art Resource, New York.© by Artists Rights Society, New York/VG Bild-Kunst, Bonn.
22. Magdalena Droste, *Bauhaus 1919–1933* (Köln: Taschen, 1988), 58.
23. Ibid., statement by László Moholy-Nagy.
24. M. Miserocchi, "Wie wohnt Mussolini," *die neue linie* (March 1930), 10–11.
25. For the issues September 1929, February, June and December 1930, May and December 1931, July and October 1932, January and May 1933.
26. For a more detailed discussion of the journal *die neue linie* and Moholy-Nagy's designs, see Anna Wessely, "La moda della politica? Della vulnerabilità dell'arte e della lunga immunità della moda nei confronti della politica," in Bernardo Valli et al. (eds.), *Discipline della moda. L'etica dell'apparenza* (Napoli: Liguori, 2003), 215–229.
27. Magdalena Droste, "László Moholy-Nagy—Zur Rezeption seiner Kunst in der Weimarer Republik," in Georg Jäger and Gudrun Wessing (eds.), *Über Moholy-Nagy*, 26.
28. According to Xanti Schawinsky "Moholy helped people who were persecuted and in the most dangerous situations sheltered some of them in a room adjoining his studio in Kaiserdamm." Quoted from Passuth, *Moholy-Nagy*, 401.
29. He produced publicity materials for the Jenaer Glaswerke in 1933–1939 and visited Jena several times in this period, see Gudrun Wessing, *Als Lichtvisionen von Schatten überholt wurden*, 45–72.
30. Letter to Hepworth and Nicholson, quoted by Wessing, *Als Lichtvisionen von Schatten überholt wurden*, 57–58.
31. Quoted from Passuth, *Moholy-Nagy*, 405.
32. These are Moholy-Nagy's expressions in a letter to his brother from Chicago, dated August 27, 1937. English translation in Passuth, *Moholy-Nagy*, 406.
33. Being a cosmopolitan nomad does not exclude part-time activity as a political exile. The chances a victory of the Allies over fascism might open for a democratic renewal of Hungary mobilized the Hungarian exiles all over the world. In September 1941, the various local clubs and associations of Hungarians in America suspended their quarrels and created the Democratic Alliance of Hungarians in America, electing Oscar Jászi as their president. The Alliance worked to achieve permission for Mihály Károlyi (Prime Minister of Hungary after the revolution of October 1918 until the Communist takeover in March 1919) to enter the United States in order to mobilize Hungarians in America as well as to win the State Department for his program for a democratic Hungary after the war. The Chicago group was dissatisfied with the meagre results of these efforts and decided to leave the alliance. In a long letter to Jászi, Moholy-Nagy explained their political position ("all of us stand on the principles of continuing and developing [the goals of] October and have regarded Károlyi as our leader") and their reasons for leaving the alliance (they demanded "action by the Hungarian speaking masses" instead of the futile diplomatic negotiations Jászi was conducting). Moholy-Nagy's letter, dated September 13, 1942, was published by Lajos Varga in *Múltunk* 45, 3 (2000): 231–235.
34. On the history of the three successive design schools organized by Moholy-Nagy in Chicago, namely The New Bauhaus, the School of Design, and the Institute of Design,

see Alain Findeli, *Le Bauhaus de Chicago. L'œuvre pédagogique de László Moholy-Nagy* (Québec: Septentrion, 1995).

35. Victor Margolin, *The Struggle for Utopia. Rodchenko, Lissitzky, Moholy-Nagy 1917–1946.* (Chicago, London: University of Chicago Press, 1997), esp. ch. 6: "Design for Business or Design for Life? Moholy-Nagy 1937–1946," 215–250.

36. *The New York Times* (12. 9. 1937). The article is reprinted in the catalogue *50 Jahre new bauhaus. Bauhausnachfolge in Chicago* (Berlin: Bauhaus-Archiv, 1987), 54.

Chapter Seven

Occult Encounters and "Structural Misunderstandings" in Exile: The Surrealists and the *Institut für Sozialforschung* in the United States

Laurent Jeanpierre

There is a fact[1] little attended in the otherwise abundant chronicles of May 1968 in France: an exhibit of works by surrealists in exile in the United States during World War II was held in Paris at the American Center during the so-called events.[2] Meanwhile, Parisian walls were being covered with slogans of surrealist inspiration only two years after André Breton's death, and the surrealist refrain of desire and of imagination in power was often heard in the marches and the general assemblies. While surrealist artists gathered to commemorate the past, only a few of them were enthusiastically welcoming these new social movements. Yet, one year after the political crisis of May 1968 and after more than fifty years of activity, the surrealist group of Paris chose to dissolve.[3] The German philosophers and social scientists of the *Institut für Sozialforschung* were caught in an equally paradoxical bind during the same period. While many radical students in Germany and the United States claimed the Frankfurt School's writings fostered and justified their actions, they also verbally and physically protested against what they saw as the "resignation" of Adorno and Horkheimer, which contrasted with the mounting protests in the world. In response, the two former exiles republished in 1969 the major Frankfurt School work of the period of World War II, *Die Dialektik der Aufklärung*. It included a new introduction that undermined the use most of the students were making of it, and it castigated them harshly for politically instrumentalizing the critique of instrumental reason developed in the book. In France and Germany during the 1960s–1970s, the World War II exiles of surrealism and Critical Theory[4] appeared as a covert presence and a *contested legacy*.

The purpose of this piece is neither to uncover the reasons why this recent critical period saw a return of these two critical traditions—a return, we claim, of the socially repressed before 1968—nor to reflect on the original protagonists' experience of exile in the United States. Yet it is worth noticing that despite the rapid growth of the historiography of both surrealism and Critical Theory after 1968, the studies dealing with intellectual relationships and possible interactions between the two movements are still very few.[5] Adorno and Horkheimer had been interested in the surrealist movement since the 1920s, but the reverse relationship never developed. In the mid-1930s, both movements independently became critical of the orthodox communist and psychoanalytical traditions, while remaining faithful to Marxism and indebted to Freud. They shared similar views on the critical function of culture in society and a common interest in art and daily life as sites of political resistance against capitalism. The surrealist artist and writers and the philosophers and social scientists of the Frankfurt School represented archetypes of the critical and public intellectual. They held similar "avant-gardist" positions in their respective national intellectual fields.[6]

Yet, regardless of the homologies of their social positions and the affinities of their political standpoints, the two groups never met during their stay in exile in the United States. This failed encounter in America could certainly be explained by contingencies: Adorno and Horkheimer moved to California after 1940, while Breton and his friends remained in New York and its surroundings. Moreover, the absence of direct interactions between the surrealists and the German radical philosophers in exile may be due to some more structural and institutional elements of the period's intellectual life. Above all, the artistic and philosophical fields were less connected in the United States in the 1940s that they had been in the Weimar Republic or in Paris in the 1920s and 1930s. After World War II, the leading figures of the two groups and many others decided to return to Europe. They remained deeply affected by their stay in the United States and in some important measure oriented to it, not least in their respective pessimistic views on modernity. During the Cold War for example, they vigorously criticized American domination and ideology. In other words, surrealist artists as well as philosophers from the *Institut für Sozialforschung* remained "permanent exiles," long after the war period.

By insisting here on this apparent affinity between the two movements in exile and after, I suggest how fruitful it could be to engage in some comparison between different national groups of exiles and this way to call for a *de-nationalization of exile studies*. I imply here that the differences in the discourses of André Breton and of Theodor Adorno and Max Horkheimer[7] while in exile in the United States can be understood by taking into account the differences in the contribution of each group to the *internationalization of intellectual fields*. I eventually suggest that they also express the struggles in the "world-system" of symbolic production between French and German intellectuals to impose their respective national definition of intellectual commitment.

Some "Structural Misunderstandings"

During their American stay both groups were occupied by the theme of myth and its social and political meaning. The surrealists, having left the Communist Party in 1935, had already been discussing before 1940 the social and political need for what they called a "new myth."[8] Exiled in New York beginning in the summer of 1941, Breton repeatedly called, in a number of his publications, for "the development of a *collective myth* for our times."[9] According to him and his friends, this would be the only way to overcome the crisis of rationalism in which the Western world was caught up, as evidenced by Nazism and the World War. As for the *Dialectics of Enlightenment* of Adorno and Horkheimer, which was published for the first time in 1944, it represents itself as an anthropological study of the relationship between myth and Enlightenment rationality in the history of Western thought. "Just as the myths already comprehend Enlightenment," Adorno and Horkheimer summarized in the beginning of their essay, "so Enlightenment with every step becomes more engulfed in mythology."[10] It is this paradoxical historical dialectic, they contended, that explains the historical possibility of Nazism and indeed modernity in general, manifested as the dual advent of an administered society and its *Doppelgänger*, a regime of violence as radical as it was banal.

In the history of ideas, these respective positions on myths and their social function have been commonly presented as radically antithetical. According to Jürgen Habermas and other continuators of the Frankfurt School for example, the surrealists have been guilty of mere subjectivism since the interwar years and even more so during the last world war and of a nostalgia for an uncontaminated nature. On the contrary Critical Theory has delved seriously into the interweavings of rationality and irrationalism that had led to Nazism. In fact, as early as 1937, Adorno and Horkheimer had already decided, often against the opinion of Benjamin, that the surrealist attempts to seek new myths was utterly Romantic and that, as such, it could be interpreted as proto-fascist in spirit.[11] Because of this, in Germany the surrealists can also be supposed to be guilty of complicity in the neo-Romantic German *Zeitgeist* that had gradually led the most well intentioned minds of Weimar Germany—from Stefan George's circle to Expressionism—to make Nazi ideology possible and legitimate. In this dominant interpretation, the supposed opposition between surrealism and Critical Theory in the 1940s concerning the question of myths is almost always reduced to the supposed antithesis between a flight into irrationalism and a fidelity to a rationalism waiting to be developed. This interpretation by Adorno and Horkheimer, as well as by the recognized heirs of their legacy, is due to their position in the German intellectual field since the 1930s, and their consequent need to fight equally against anything that might grant credibility to either mythology or to positivistic science. Like the decision of French heterodox Marxists of the 1950s to import Heidegger instead of Adorno, this outcome belongs to the long history of "structural misunderstandings" between French and German intellectuals in the twentieth century.[12]

"On the Survival of Certain Myths . . ."

But a more careful look at the surrealists' activities while in exile in the United States reveals a need to revise this constructed mythology about the surrealist quest for a new myth. For the catalogue of the New York exhibit "First Papers of Surrealism" organized by Marcel Duchamp in 1942, Breton created a collage entitled "On the Survival of Certain Myths and on Some Other Myths in Growth or Formation."[13] Each page of this collage was illustrated by two images as well as by quotations from diverse mythical sources. Breton chose to deal with the following myths: the Golden

Age, Orpheus, Original Sin, Icarus, the philosopher's stone, the Grail, artificial man, interplanetary communication, the Messiah, regicide, soul mates, triumphant science, Rimbaud, and the *Übermensch*. This collage's techniques, its irony, aimed at undermining any attachment of affects to already existent myths. Science, and with it the ideology of progress, was derided for example with a quote from Alfred Jarry; the figure of Nietzsche evoking the *Übermensch* was combined with the new comic book icon, *Superman*. Breton's arrangement thus served to question the effectiveness of hackneyed myths, and especially those that made up Nazi ideology. Speaking to students at Yale in 1942, Breton explained his approach: he favors a "practical preparation for an intervention in mythical life that would first, on the largest scale, take the form of a *cleansing*."[14] His collage was therefore to work as a *demystifying* mechanism. Nothing allows us to cast doubt on Breton's dissatisfaction with either Romantic mythologies or contemporaneous mythologies, while he was in exile in the United States.

Nazism, he declared to the Yale students, had produced "myths that are incompatible with the harmonious development of mankind."[15] But at the same time, Breton would state after the war's end that to refuse any outlet for mythical thought alongside rational thought would "mean making it destructive, leading it to explode within the rational domain and to obliterate it (delirious cult of the leader, two-bit messianism, etc.)."[16] The argument is very close here to the dialectic of myth and reason outlined by Adorno and Horkheimer in their 1944 book. In fact, for Breton and the surrealists in the 1940s, the call to search for a modern myth implied a redefinition of the myth, a sociological rereading of its function in history. Breton also supported Marx's idea that the function of myth is to disclose the role of natural forces in the imagination. Myth for surrealists is not a false idea—blind faith or ideology. What surrealists called "myth" must contribute, Breton wrote in 1942, to "making connections among human beings fruitful and desirable once more." It should help rebuild social ties broken by war and modernity, reestablish social bonds in a *new relationship to nature* through the "most universal contact, without prejudice among living beings."[17] Moreover, this modern myth Breton longed for was not to be carried by a single individual—as art, poets and artists have often done—but by a new collectivity.

"The world is dying from rationalism, from *closed rationalism*," Breton affirmed in 1941, "physical violence is unconsciously accepted, justified as the outcome of mental passivity."[18] When Breton during World War II turned to the question of occult beliefs, we understand that it was in order to found *an open rationalism*, a venture that was complementary to attempts to create a non-regressive myth. That is why when referring to the surrealist project in general, Gaston Bachelard can call it a "surrationalism." It also means that far from being diametrically opposed, as they often are by most commentators, the ideas of Breton, Adorno, and Horkheimer while in exile have been dealing with the very same problem.

Toward a New Dialectic?

At the heart of their philosophies, each of the two groups first substituted the man/nature dialectic for the Marxist historical hypothesis of class struggle. Adorno and Horkheimer sought a recognition of nature in man that would also be a new relationship of man to nature—beyond the former's domination and the latter's submission.

> Every attempt to break the natural thralldom by breaking nature itself enters all the more deeply into that natural enslavement [. . .]. Men always have had to choose between their subjection to nature and the subjection of nature to the self.[19]

The following excerpt from Breton's "Situation of Surrealism between the Two Wars" echoes the previous quote almost word for word:

> Among the contradictions that are fatal to us, the one whose resolution requires the greatest ambition—the one to which I personally devoted most of my effort—is the one that brings man and nature into conflict from a human standpoint, since man perceives the necessity at work in nature and his own necessity to be at odds with each other.[20]

Surrealism and Critical Theory alike sought in the *mimesis* of the animal world, childhood, magic, poetry or works of art, an alternative to the unreasoned march of instrumental or calculating reason, which is nothing else, for Adorno and Horkheimer, than the "mimesis unto death." The surrealists also rejected what the German philosophers would have called the monopoly of instrumental reason in the relationship between man and nature. The old Jesuit precept that "the end justifies the means" should be abolished, Breton wrote, "Only by working to abolish this precept, which has gained monstrous vigor at the entrance of the inquisitorial den, can we claim to serve the cause of freedom—and *not* by vainly reducing freedom to the arbitrary faculty of doing as we please."[21]

It is now interesting to return to the situation in the 1960s and 1970s when these debates between surrealism and Critical Theory returned, I believe, in concealed fashion. In the *Dialectics of Enlightenment*, the mythical ideologization of the past, the rationalized domination of nature and the social order produced by capitalist exploitation are all tied to a common origin. Adorno and Horkheimer participate here paradoxically in something they constantly criticized in mythical thought: the fact that it expresses the passion for origins that has always accompanied the feeling of modernity. They often fall back on the "Master Narrative" of this modernity reversed, the story of the "eclipse of reason," derided by Lukàcs as coming from the "Grand Hotel Abyss." Their often-apocalyptic pessimism explains in large part the accusations made against Critical Theory by the radical fringes of the German student movements in the 1960s–1970s.

One of the common criticisms of surrealist art after 1940 was its "recovery" by capitalism and the American art world; for some, the European avant-garde had been integrated into the culture industry.[22] In other words, once in exile in the United States, surrealism became merely a forerunner of what the situationist Guy Debord called in 1967 the "society of spectacle," a society where mythology has become materialized. Surrealism, some said after World War II and in the 1970s, had resigned itself to making art into simple fetishistic commodity rather than a device for creating new passions outside of the realm of art for art's sake.

Like the attacks that sought to counterpose the supposed irrationalism of surrealism against the rationalism of Critical Theory, these attacks probably miss the *dialectical correspondence* between the German avant-garde philosophers and the French avant-garde writers and artists. This correspondence was not restricted to the 1930s but rather continued after 1940. Elisabeth Lenk, who played a prominent postwar role between the two groups, even considers that surrealism was the practical side that the Critical Theory was lacking, and Critical Theory was the theory surrealist art was missing.[23]

Situations of Exile and Forms of Internationalization

In order to explain this dialectical relation between the two movements one should now come to the social situation of exile as both groups experienced it. Remarkably, they faced the same difficulties and expressed the same reticence when confronted with American culture. For example, Breton notoriously refused while in New York to speak English and requested that the doorkeeper of the Museum of Modern Art greet him in French. Adorno complained of the blind empiricism of American social sciences,[24] and pointed to the ideological danger of popular culture. Politically also, both groups were highly skeptical of the Allies' war effort and looked forward to the reconstruction period with pessimism. Yet recent studies have shown that such historiographical *clichés* need to be revised: surrealists in fact established connections with the artistic Bohemia of New York; and Adorno and Horkheimer were indifferent to neither the movie industry nor the American social sciences.[25]

Yet, it is fair to say that during their stay in America the surrealists contributed more to the internationalization of art and literature than the members of the Frankfurt School did for philosophy and the social sciences. Although both groups seem to share the rejection of American values and American norms of cultural production, it appears that the reception of the surrealist movement and the Frankfurt School in the United States in the 1940s differed greatly.[26] In the international space of artistic and literary exchanges the surrealist could occupy a prominent position since Paris was at the center of the World Republic of Letters before 1940.[27] At the same moment in the philosophical field the process of internationalization is not as developed. Even though German philosophy was a strong presence, especially in the nineteenth century, the United States also developed a tradition of its own, and national traditions in philosophy compete with each other more openly than in the artistic fields. Moreover, surrealists benefited from the emergence of a national market for modern art in New York. Fringes of the cosmopolitan bourgeoisie of the East Coast and a few newcomers in the artistic and literary fields had interests in promoting their work.[28] Surrealist exiles were constrained in return to legitimate American avant-gardist endeavors. On the contrary, no American intellectual groups had interests in promoting the ideas of Adorno and Horkheimer who remained isolated. It may be a reason why their universalism became more abstract—that is, without any material and social base—than the surrealist one:[29] whereas the surrealists could criticize the abstract universalism of European intellectuals by calling for a new collective bond that would produce a new collective myth, Adorno and Horkheimer could only refer to the values they had already held before their exile.

Moreover the intellectual socialization and the social composition of both groups had been very different prior to their exile. The weight of *Bildung*, of traditional bourgeois culture and the attachment to one's national culture and discipline was more important for the main philosophers of the Frankfurt School than it was for the surrealist artists and painters.[30] Whereas the latter were coming from various social backgrounds and entered the artistic field without specific or professionalized training,[31] the former had been educated to represent the German philosophical tradition and culture.[32] Recent historiography has confirmed this view by showing how Adorno and the Frankfurt School also played a key role in the public debates reflecting upon the political architecture of postwar Federal Germany.[33] In the French case, the reference to one's national identity works completely for the surrealists. Marginalized from the start in their own professional field—whether literary or artistic—and marginalized also as artists or writers in an intellectual field where professors were becoming dominant, the surrealists never had the chance nor the interest to invest national culture or to be indebted to it. After having experienced World War I, the founders of surrealism and Breton first among them rejected vigorously any national stance. In exile he and his friends resisted to the renationalization of political and cultural activities. They did not for example enrol for this precise reason in the de Gaullist movement. On the contrary the universalism of Adorno and Horkheimer during the same period may well have been hiding some deeper faithfulness to national traditions.

Conclusion

With the surrealist theme of the creation of a new myth, artists, writers, scholars were called on during the critical times of the World War II to abandon their elevated positions, their sanctuary, to leave their university chairs and studios behind, in order to participate in the experimental and collective creation of new "forms of life." Confronted in exile with the growing heteronomy of their practice the surrealists did not try to defend its autonomy abstractly by retracting into a universalistic defense of art for art's sake. For them, the task of the artist and more generally the intellectual was no longer a question of saving the world through poetic re-enchantment but rather of building a new form of being together that could be extended. Breton and the surrealists in exile actually proposed a movement beyond the aesthetic through ethics.

Thus the major difference between surrealism and Critical Theory after 1940 lies in the relationship they seek to establish between *subject* and *truth*. Following the tradition of the early age of ancient philosophy, the surrealist approach is aimed at *good life*. Critical Theory remains bound to truth in a relationship of knowledge, even when critical. Thus in the continuing dialogue between surrealism and Critical Theory, one last specter may perhaps be seen, the specter of a new ethical alliance that Gramsci once spoke of: between the "pessimism of reason" and the "optimism of the will."

Notes

1. I would like to thank Ames Hodges for his help in translating the French elements of this text.
2. Nelcya Delanoë, *Le Raspail Vert: l' American Center à Paris, 1934–1994. Une histoire des avant-gardes franco-américaines* (Paris: Seghers, 1994), 138.

3. On the split of the surrealist group in 1969, see Carole Reynaud-Paligot, *Parcours politique des surréalistes, 1919–1969* (Paris: CNRS, 2001), 202; Alain Joubert, *Le mouvement des surréalistes ou le fin mot de l'histoire* (Paris: Maurice Nadeau, 2001); Marie-Dominique Massoni, Conference, Prague, May 3, 2004 (to be published in *S.U.RR*).

4. By Critical Theory, I mean here the discursive productions of the group that has been recognized in historiography as the first Frankfurt School.

5. For a recent comparative approach of the two movements: "Art, raison et subversion," *Agone*, 20, coordinated by Norbert Bandier and Dietrich Hoss, 1998. See primarily: Norbert Bandier, "Écriture et action chez les surréalistes et chez Benjamin," Dietrich Hoss, "Surréalisme et théorie critique: parcours de deux courants voisins," Elisabeth Lenk, "Théorie critique et pratique surréelle," Heinz Steinert, "Pourquoi Adorno a changé d'avis sur le surréalisme après 1945?" Gérard Roche, "Affinités et inspirations surréalistes chez Benjamin."

6. On the sociological characteristics of an avant-gardist position in the literary field, see Gisèle Sapiro, "Forms of politicization in the French literary field," *Theory and Society* 32 (2003): 633–652, 645–646.

7. In this short piece I can only focus on some of the main texts written by the prominent members of both groups in exile in the United States after 1940: André Breton and Theodor Adorno and Max Horkheimer. Thus I necessarily underestimate the weight of the conflicts and the different viewpoints within each movement. For an account of the Frankfurt School in the United States, see Martin Jay, *The Dialectical Imagination: A History of the Frankfurt School and the Institute of Social Research, 1923–1950* (Boston: Little Brown, 1973). For an account of some of the activities of the surrealists in the United States, see Martica Sawin, *Surrealism in Exile and the Beginning of the New York School* (Cambridge: MIT Press, 1995); Romy Golan, "On the Passage of a few Persons Through a Rather Brief Period of Time," in Stephanie Barron (ed.), *Exiles + Emigrés. The Flight of European Artists from Hitler* (Los Angeles: Museum of Contemporary Art, 1997): 128–146; Jacqueline Chénieux-Gendron "Surrealists in Exile: Another Kind of Resistance," in Susan Rubin Suleiman (ed.), *Exile and Creativity, Signposts, Travelers, Outsiders, Backward Glances* (Durham, London: Duke University Press, 1998): 163–179; Laurent Jeanpierre, *Des hommes entre plusieurs mondes. Intellectuels français réfugiés aux États-Unis pendant la Deuxième guerre mondiale* EHESS: Actualités 2005 <http://www.actualités.ehess.fr/nouvelle488>.

8. For an overview of these debates on myth and their political meaning see: Anette Tamulty, *Le surréalisme et le mythe* (New York, Paris: Peter Lang, 1995); Jacqueline Chénieux-Gendron, Yves Vadé (eds.), *Pensée mythique et surréalisme* (Lausanne: Lachenal & Ritter, 1996); Didier Ottinger, *Surréalisme et mythologie moderne. Les voies du labyrinthe d'Ariane à Fantômas* (Paris: Gallimard, 2002).

9. André Breton, "Situation of Surrealism Between the Two Wars," *VVV* 2–3 (March 1943), in André Breton, *Free Rein (Le Clé des Champs)* (Lincoln: University of Nebraska Press, 1995) (hereafter cited as *Situation of Surrealism*).

10. Theodor W. Adorno und Max Horkheimer, *Dialectic of Enlightenment*, transl. John Cumming (New York: Continuum, 1996 [1944]): 11–12 (hereafter cited as *Dialectic of Enlightenment*).

11. Theodor Adorno, Walter Benjamin, *The Complete Correspondence, 1928–1940* (Cambridge/MA: Harvard University Press, 1999).

12. On this history and its explanation by this notion of "structural misunderstanding" see Bourdieu, 1999: 221.

13. André Breton, "On the Survival of Certain Myths and on Some Other Myths in Growth or in Formation," *First Papers of Surrealism* (Coordinating Council of French Relief Societies, 1942).

14. Breton, *Situation of Surrealism*, 66–67. My emphasis added.

15. Breton, *Situation of Surrealism*, 52.
16. Ibid.
17. André Breton, "Prolegomena to a Third Manifesto of Surrealism or Else," *VVV* 1 (June 1942): 18–26, in André Breton, *Manifestoes of Surrealism* (Ann Arbor: University of Michigan Press, 1967), 294. Emphasis mine.
18. André Breton, "Interview with Charles-Henri Ford," *View* (October–november 1941): 7–8, in André Breton, *Conversations (Entretiens)* (New York: Paragon House, 1993).
19. Adorno, Horkheimer, *Dialectic of Enlightenment*, 13, 32.
20. Breton, *Situation of Surrealism*, 65.
21. André Breton, "Interview with Jean Duché, 5 october 1946," in André Breton, *Conversations (Entretiens)* (New York: Paragon House, 1993), 205.
22. Among many others, see: Jules-François Dupuis and Raoul Vaneigem, *Histoire désinvolte du surréalisme* (Nonville: Paul Vermont, 1977).
23. Lenk in Agone, 1998.
24. As in As in Theodor W. Adorno, "Scientific Experiences of a European Scholar in America," in Bernard Baylin and Donald Fleming (eds.), *The Intellectual Migration, Europe and America, 1930–1960* (Cambridge: The Belknap Press of Harvard University Press, 1969), 338–370.
25. See for example Thomas Wheatland's contribution in this volume.
26. Lewis Coser, *Refugee Scholars in America. Their Impact and their Experiences* (Newhaven: Yale University Press, 1984).
27. Pascale Casanova, *La République mondiale des Lettres* (Paris: Le Seuil, 1999).
28. On the interests of various American groups, inside and outside the art world, to promote surrealism see Serge Guilbaut, *How New York Stole the Idea of Modern Art*, trans. Arthur C. Goldhammer (Chicago: University of Chicago Press, 1983).
29. On the general relation between social interactions of different national intellectuals and the quality of their universalism in situations of exile see the remarks of Hans Speier, "The Social Conditions of the Intellectual Exile," *Social Research* (1937): 316–328.
30. On German Bildung and its weight in the intellectual life since the XIXth Century, see Fritz Ringer, *The Decline of the German Mandarins: The German Academic Community, 1890–1933* (Cambridge: Cambridge University Press, 1969); Reinhard Koselleck, *Bildungsbürgertum in 19. Jahrhundert*, vol. 3, "Bildungsgüter und Bildungswissen" (Stuttgart: Klett-Cotta, 1990); Bernhard Giesen, *Die Intellektuellen und die Nation. Eine deutsche Achsenzeit* (Frankfurt: Suhrkamp, 1993); Dominic Boyer, *Spirit and System: Studies of Media, Intellectuals, and Dialectical Social Knowledge in Modern German Intellectual Culture* (Forthcoming).
31. Norbert Bandier, *Sociologie du surréalisme, 1924–1929* (Paris: La Dispute, 2000).
32. For a developed sociological comparison between the structures of the German intellectual field and the French intellectual field, see Christophe Charle, *Les Intellectuels en Europe au XIXème siècle. Essai d'histoire comparée* (Paris: Le Seuil, 1996).
33. See for example Alfons Söllner's contribution in this volume.

CHAPTER EIGHT
THE DAVOS DEBATE, SCIENCE, AND THE VIOLENCE OF INTERPRETATION: PANOFSKY, HEIDEGGER, AND CASSIRER ON THE POLITICS OF HISTORY

Gregory B. Moynahan

In 1939, six years after his initial exile from Nazi Germany, and three years into a stay in Sweden that was to end in 1941 with a final move to America, the philosopher Ernst Cassirer wrote a full-length study of the Swedish political and legal theorist Axel Hagerstrom. That Cassirer, one of the leading German philosophers of his day, would at the age sixty-eight not only learn Swedish, but also apply his knowledge to a full-length study of a leading Swedish legal philosopher was entirely typical of his stay not only in Sweden, but of his previous exile in England and his final stay in America. More than almost any other émigré figure, Cassirer displayed a remarkable ability to blend with his adopted countries. Yet in the introduction to this book, Cassirer explains that this ability was not based on his intellectual proximity to his adopted countries. Rather, he emphasizes his extreme distance from them. He begins his book by situating his own position with the following quotation from Voltaire's *Letters from England*. "A Frenchmen who arrives in London," Voltaire writes, "finds things very different in philosophy as in everything else. He has left the world full; he finds it empty. In Paris they see the universe as composed of vortices of subtle matter, in London they see nothing of the kind. For us, it is the pressure of the moon which causes the tides of the sea, for the English, it is the sea, that gravitates against the moon [. . .]. The very essence of things has totally changed. A common understanding exists neither over the definition of the soul nor of matter."[1] The problem of philosophical translation between Germany and neighboring Sweden, Cassirer argues, is so great they would constitute nearly different worlds, even with their present politics aside. Exile philosophers are to immerse themselves in a foreign tradition, to reread it from within, and in this way to bring it into a new dialog with their own position, while at the same time leaving behind documents that may in the future be read as a meaningful synthesis of nearly incommensurable intellectual positions.[2]

Cassirer's ironic quote from Voltaire also hints at an equally important earlier "translation" of his German works in relation to natural science. By reminding his readers that philosophy used to be coexistensive with natural philosophy, Cassirer recalls earlier transformations in his own work. Even before exile, Cassirer had undergone a shift from a world in which his assumptions could be shared and communicated to one in which they could not. This shift had indeed involved his understanding of "the very essence of things" as they are reflected in natural science. Cassirer's early work was grounded on a philosophy of natural science associated with the Marburg school of his teacher Hermann Cohen, and he had staked a great deal on the popular reception of this new vision of natural science and its capability for social transformation.[3] Following World War I, however, the very tenability and relevance of a philosophy of science became widely questioned, even as new forms of "life-philosophy" and ontological philosophy began to be celebrated on the Weimar philosophical scene.

Before Cassirer translated his works and assumptions spatially in exile, in short, his works had already been translated across a temporal and generational divide caused by the rupture in understandings about natural science. His tactics in exile were closely related to his tactics for dealing with all forms of philosophical, social and political difference: even as he advanced his own philosophical perspective, he strongly believed in always modulating it for his immediate audience while leaving clues for its future "reconstruction" as a complete whole. Understanding not only Cassirer's philosophy but the era in which it developed depends on retracing this double translation.

In this chapter, I attempt to recast the meaning of Cassirer's early work and reception by developing the implications of his writings up to his 1928 "Davos debate" with Martin Heidegger. These, I argue, are central for understanding Cassirer's politics as it developed from the Wilhelmine period into the more explicitly political work of the crisis and exile years. As a means of clarifying Cassirer's "double translation" out of his earlier philosophy and into exile, I use the early reflections upon the Davos debate by one of Cassirer's most adept and influential students, the art historian Erwin Panofsky, who significantly shaped postwar North American, and ultimately European, art history. I show that for Cassirer and Panofsky the key importance of the issues raised by the Davos debate and its associated texts lay in establishing the basis for a new methodology of historical study and a means of minimizing "interpretative violence" of the sort Cassirer found in Heidegger's ontology. Far from being solely the introductory forum for Heidegger's ontology, as is often claimed by historians of philosophy, the Davos debate also displayed a critical and phenomenological historical method of great consequence.

Reading Cassirer's work in light of both transformations, those of his early work on natural science and his experience of exile, is crucial to understanding the political meaning of Cassirer's philosophy. Following his death in 1945, Cassirer was widely received as the epitome of the "good humanist" by his history students and as a "rationalist" and "Enlightenment thinker" by those in the philosophy of science. Both positions, I suggest, readily collapsed into a misrepresentation of Cassirer's actual position. Moreover, they encouraged a misreading of the Davos debate by proponents of both sides as a confrontation that articulated neo-Romantic hostility to the supposed "intellectualism" and inorganic nature of modern thought or as a conflict between the stormy ideals of Heidegger's youthful cohort and the genteel ideals of Cassirer's bourgeois generation.

In opposition to such readings, whether in praise or blame, I suggest that Cassirer clearly saw that a defense of Enlightenment ideals required a radical reformulation of the Western philosophical tradition, and that this reformulation confounded the familiar stereotypes. In particular, for Cassirer the ideal of Enlightenment required first and foremost a turn to skepticism, self-questioning, and to new methodologies that allowed this continual reflection. Ironically, Cassirer found in Heidegger's path-breaking work the culmination of an intellectual tendency toward claims of absolute knowledge and essentialism that was already present on the side of Enlightenment that Cassirer rejected and that took exaggerated form in Romanticism and its later appearances as "life-philosophy."

The debate between Cassirer and Heidegger will thus be read from Cassirers' perspective as a conflict over the nature of Enlightenment and its Romantic successor movements, both intellectually and politically. The product of that encounter yielded an essential intellectual resource that eventually assisted key figures of the intellectual emigration, directly or indirectly, in making a transition from the predominately anti-Enlightenment mental climate of the Weimar Years to the American intellectual culture premised on unproblematic celebration of Enlightenment. Precisely through this transition, however, the original meaning of Cassirer's work and the value of his debate with Heidegger were fundamentally distorted. The object of the present study is to recover invaluable aspects of Cassirer's thought, freed of the simplifications introduced by unsubtle appropriations of his "double-translation" strategy in exile.

The Crisis of the Sciences and the Counterposed Positions at Davos

The opacity of the Davos debate arises in part from its summary nature: both philosophers are forced to make short telegraphic points to an audience that understands neither their affinities in philosophical training nor their profound disagreements on key issues. Based on a wider debate over Kantian philosophy spurred by Heidegger's recent *Kant and the Problem of Metaphysics* (*Kant und das Problem der Metaphysik*) (1929) and Cassirer's review of it, Heidegger and Cassirer each sees crucial flaws in the other's positions, which he imagines are nearly invisible to his opponent. Heidegger views Cassirer's supposed failure to turn from the problem of the ontic or particular sciences to ontology as blocking a true philosophical hermeneutics and trapping Cassirer in an outdated definition of Enlightenment rationality and humanism. Cassirer believes that Heidegger's new ontology is, ironically, not hermeneutic enough, since precisely the hidden essentialism of his ontology blocks the path of a *historical* hermeneutics that recognizes philosophy's embeddedness in language and ontic history—and thus blocks a path to Enlightenment skepticism. The enduring value of the Davos debate despite its summary nature is that it forces us to read each philosopher's position in relation to the other's, and thus to try to develop both a philosophical and historical hermeneutics. As we will see, Panofsky's work represents one early attempt to do precisely this.

Cassirer, as commentators such as Jürgen Habermas have noted, makes a version of what will later be described as the "linguistic turn" argument against the new ontological philosophies.[4] Habermas and other commentators have underestimated the radicalness of Cassirer's linguistic turn, however, since it is rooted in Cassirer's earliest work on science, which uses not language, but the group theory of mathematics to create a sort of universal semiotic that claimed that all reality—including what had been known as "subject" and "object"—were radically plastic, relational, and historical.[5] Precisely this plasticity meant that neither "commonsense" rationalism nor traditional humanism were inherent in Cassirer's philosophy. Starting with his *Freedom and Form* (*Freiheit und Form*) of 1916, Cassirer would argue that the development of political, social, and cultural forms had to be careful nurtured and criticized, since these fragile forms create different forms of subjectivity, objectivity, and meaning—and thus different political environments.[6] Cassirer's postwar project was to find a further grounding for form and to work out its hermeneutic paradoxes through a study of the phenomenology of symbols, and thus to create a fully immanent and relational definition of experience—as most clearly occurs in his much misunderstood *Philosophy of Symbolic Forms*.

The Davos debate can best be introduced by Cassirer and Heidegger's similar reading of the so-called crisis of sciences that had gradually developed in Europe in the previous thirty years. Curiously, neither Heidegger nor Cassirer's questions the reality of the problem, even as their decisive differences reflect their differing responses to the problems of Enlightenment. If we take Heidegger's own definition of the "crisis" of the sciences from *Being and Time*, we find that it is similar to Cassirer's more extended definition of the crisis and understanding of the contemporary state of science as represented in texts such as *The Problem of Knowledge, Determinacy and Indeterminacy in Modern Thought*, and *The Philosophy of Symbolic Forms*. As both would probably note, the similarity is based on their common starting point in the so-called Marburg School of neo-Kantianism that centered on the figure of Hermann Cohen.[7] Nonetheless, it is also suggestive—as the preeminence of the Davos debates in European intellectual history itself suggests—of a common vanguard position on the topic in Europe in the late 1920s that was perhaps the most important remaining salient of the Marburg school.

In *Being and Time* Heidegger introduces his understanding of the crisis of the specialized—or, as he calls them, the "ontic"—sciences as follows: "A science's level of development is determined by the extent to which it is capable of a crisis in its basic concepts. In these immanent crises of the science the relation of positive questioning to the matter in question becomes unstable.[8] Heidegger cites mathematics, physics, biology, history, and theology as *Wissenschaften* in which the basic concepts (roughly number, motion, life, time, and faith) have become "capable of crisis."[9] Cassirer held to a similar understanding of a crisis of modernity extending across both the natural sciences and humanities. Both agreed that the nature of the "crisis" in the sciences was that no particular science could any longer present a static definition of individuals, society, or the world. The modern project of "rationally" defining the individual, society, and the world in terms of absolute structures had thus collapsed. Both further agreed that the dichotomy of nature and culture was inadequate to address the crisis.

For Cassirer, human culture produces unlimited fields of study (or perception) over time and corresponding new objects (and subjects). These fields can be incommensurable yet legitimate—there can be both drastic historical shifts in meaning (paradigm shifts) as well as fundamental differences in the realities of both different societies and different sciences. Human being is defined by an unlimited plasticity—much as in the Renaissance figure, Pico della Mirandola, there is in fact no "human nature" other than this plasticity—which nonetheless has a common horizon of structure in action. Correspondingly, there is an infinite capacity for creative development over time despite the continuous appearance of absolute limitation in the moment. Indeed, an education in the possibility and (as against Hegel) open or contingent quality of this dialectic—however static present life may appear—was central to Cassirer's philosophy. The active discernment of a unifying sense that links this infinite field can only be found by specific and local studies that suggest the play of given and possible, ontic and ontological. This is the core of Cassirer's reinterpretation of the Enlightenment as a political project.

For Heidegger, in contrast, such a project risks losing itself in the varied ontic projections of human being. "The genuine principle of order has its own content which is never found in ordering, but is rather already presupposed in ordering [. . .]. And if the 'world' itself is constitutive of *Dasein* [Being, lit. "being there"], the conceptual development of the phenomenon of the world requires an insight into the fundamental structures of *Dasein*."[10] The key to understanding existence is not the unlimited production of knowledge, objects, or perceptions in ontic science, but the underlying foundational structure (ontology) that underlies all such perceptions, what he termed the "structure of care." The nature of *Dasein*, he maintained, is that it comes to understand itself through its particular projects. This is critical for the crisis of sciences since *Dasein* frequently confuses itself with its specific ontic endeavors; in short, humans come to understand themselves through their sciences, but only in the limited manner characteristic of objects. It is characteristic of *Dasein* then, that it confuses its ontological nature with its particular ontic projects: it is thus what Heidegger calls, introducing a concept critical in the subsequent argument, an "ontic-ontological being."[11] The crisis of the sciences can never be solved by the ontic sciences themselves, since the crisis of the sciences reflects on a deeper meaning of ontological *Dasein*, that is the underlying structure of care that is common to *Dasein's* relation to, and discovery in, all of the sciences. The overcoming of crisis in the ontic sciences becomes incidental. They are not, as in Enlightenment thinking, the prime locus of human knowledge, but merely symptoms of a deeper misunderstanding of Being.

Cassirer's Functionalist Critique of Heidegger's Ontology at Davos

As Cassirer's subsequent notes clarify, he agreed with a great deal of Heidegger's position at Davos, especially in its diagnosis of crisis. Cassirer argues, however, that Heidegger's position does not go far enough since it projects a static definition of Being onto ontological *Dasein*. This form, as Cassirer noted, has much more in common with the ancient metaphysics of substance than Heidegger suggests. In its place, Cassirer argues for a relational or, in his terminology, functional (in the loose sense of mathematical function as pure play of relations) concept, a definition that assumes that the structure of Being is only perceptible *in* concrete action. Heidegger, in response, argues that Cassirer has missed the existential basis of Being, and that his work remains a "cultural philosophy" that improperly accepts divisions in human experience and therefore remains a merely derivative ontic science.

Cassirer contended that "the essential point which distinguishes" his position from Heidegger's is a concept of Being that "starts from a variety of functional determinations and meanings" and it is this claim that at first sight appears to miss the point of Heidegger's ontology.[12] Heidegger will counter that such a functionalism simply generalizes the dessicating role of the ontic sciences, whereas his own system unveils the deeper ontological meaning at the basis of any reality. In reply, however, Cassirer argues that Heidegger is untrue to the premises of his own thought, and that he attempts to universalize knowledge in a manner consistent with precisely the worst features of the modern Enlightenment, since it provides no means for critique of its own suppositions. Cassirer's defense of the Enlightenment's valid and continuing achievements jettisons all such substantialism. He proposes, instead, a wider project of criticizing all forms of experience through attention to the immanent conditions in signs, systems of thought, and "symbolic forms."

In his 1933 essay "Le Concept de groupe et la thèorie de la perception," Cassirer spells out the implications of the functionalist argument underlying his position in Davos.[13] Developments in empirical and Gestalt psychology since 1908, Cassirer argues, confirm his earlier argument that the group of mathematical theory forms the basis for understanding transformations of structures (such as those of language, cultural forms, etc.) that affect the perception of the present moment even as they might not be directly recognized. This entire theory depends on the fact that the "subject" is never directly given in perception, but is rather only a palimpsest of the functional whole itself.[14] The only form of continuity of experience or of the self is provided by the "symbolic character" of the relation of the whole as it is expressed in the part. This ultimately is the basis of Cassirer's argument for liberal democracy: the very fragility of the "self" demands that structures—such as constitutions and the arts—are needed that insure its most productive form at a given moment.

It is important to grasp just how radical the application of this idea is. This play of relations forms a history of "problems" that is very close to what Thomas Kuhn meant by paradigms and Michel Foucault by *epistemes*.[15] It suggests both a structural reading of history and a means of what eluded structuralists in later decades—a theory of "structural change" in time. Just as the most immediate experience of "self" is based foremost on the invariances that link a changing constellation of forms, so is the basis of any historical eras self-understanding and means of acting and being similarly as follows:grounded.[16]

"For [Heidegger]," Cassirer writes, "all temporality has its roots in the 'present moment' seen in a religious sense—for it is constituted through 'care' [*Sorge*] and through the basic religious phenomenon of death—and 'anxiety' [*Angst*] (cf. Kierkegaard)."[17] Heidegger's argument was that the relation to death, or more generally to breakdown situations, allowed for a view of the ontological structures underlying all care.[18] Cassirer's central objection is that such a path, once taken, blocks any return to a Critical Theory of specific cultural or historical forms; it must remain on the level of the absolute. This is the root of what Cassirer calls Heidegger's "Averroism."[19] Through Heidegger's approach, Cassirer writes, "history as the history of culture, the history of meaning, as the life of the objective spirit is *not* thereby disclosed."[20] Ultimately, Cassirer finds the inner split epitomized in Heidegger's reading of the inauthenticity of the "They" [*das Man*] and the essentially religious and Kierkegaardian position that "Everything 'general,' all giving in to the general is for Heidegger a 'fall'—a disregarding of 'authentic' *Dasein*."[21] It is in this split that Cassirer sees his "parting of the ways" with Heidegger.[22] The consequences of Heidegger's approach are to close off the possibility of a Critical Theory of society and culture, and with them the possibility of a political critique of social forms. Ironically, it also has the function in Cassirer's view of occluding precisely the ontic-ontological horizon of meaning that Heidegger claims to clarify. This horizon in Cassirer's view occurs only as a limiting case, as an origin or *arché* in Hermann Cohen's sense, that can be traced out from the interstice between a specific ontic activity and the limit case of its ontological foundation.

The Question of Being and the Question of Art: Cassirer's Suggestion of Art as a Basis for Explaining His Philosophy

At Davos, Cassirer selects an unexpected example of what he will take as a model of "functional" analysis as alternative to Heidegger's ontology: the meaning found in art. An explication of this model with the help of Cassirer's concept of "symbolic *Prägnanz*" and Panofsky's reading of Cassirer will clarify both Cassirer's sense of the ontic-ontological horizon and the ultimate methodical stakes of Davos. Cassirer follows his statement that the fundamental difference between his philosophy and Heidegger's rests on their respective readings of being as functional or substantive with the introduction of the sphere of art as an application of philosophy where the valences of their philosophies will become manifest. After a quick snapshot of the work of his teacher Hermann Cohen, Cassirer notes that the way beyond neo-Kantianism in all its guises will not be in the direction of ontic science at all, but through a phenomenology of fields such as language and art.[23]

Cassirer's answer is not only remarkable for the manner in which it explicitly states that the problem of language will preform any description of meaning and, ultimately, ontology. Even more remarkable is that instead of focusing on spoken language, Cassirer turns suddenly to the problem of art as a model for language. This transition is particularly notable if we consider that Cassirer's earlier work had largely avoided the question of art. Yet here Cassirer takes up art as a key rhetorical and practical moment, where his theory of "objective determination," or "Objective being" in its wholeness is somehow decisive for his work's general difference from Heidegger.

In his own writings Cassirer had not yet worked extensively on the problem of art, but his student Erwin Panofsky had. Panofsky's famous study *Perspective as Symbolic Form* [*Die Perspektive als 'symbolische Form'* (1924)] had in fact explicitly used art, and particularly perspective, to define the determination of experience in symbolic form. Panofsky's piece argues that the Western tradition of perspective is neither inevitable nor objectively given, but is rather a constructed development. In keeping with Cassirer's method, the argument is thoroughly immanent, in that it is supposed to show how *only* through the historical development of the medium of signs and symbols, broadly construed, does meaning develop in the world. A symbolic form is defined by Panofsky in this context as a process where "spiritual [*geistige*] meaning is attached to a concrete, material sign and intrinsically given to this sign." Such forms exist because "there is," Panofsky writes, "a fundamental discrepancy between 'reality' and its construction." Following Cassirer, Panofsky would hold that reality is not "pre-given" or "outside" phenomenal experience; noumenal reality is an ideal vanishing point in such a manner that "matter" and "form" can only be understood as reciprocally defining each other.[24] The "fundamental discrepancy" is thus the operative condition for the form of ontic-ontological dyad that interested Cassirer and Panofsky. Perspective develops as one form of the construction of reality out of a series of historically contingent practices in a number of fields. The influences range, in Panofsky's telling, from the advent of analytic geometry to the new technique of constructing tile floors—floors whose grid suggests the technique of early perspective. Once it had appeared, however, the form of perspective effectively led to a crystallization of both an objective geometrization of space and a new definition of subjectivity, the appearance of "modern anthropocracy."

Speaking in Cassirer's presence in the Warburg library in Hamburg, Panofsky is careful to note in his lecture that the "reality," which the "symbolic form" of perspective gives us, is not in itself directly given in any manner than through its interpretation. Panofsky's point—like Cassirer's—is that *all* homogenous space, whether visual and geometric, is not the ultimate basis for establishing reality, as a naïve view of natural science might have it, but an abstraction.[25] Yet if the mathematic space of physics does not represent the "reality" of space, neither does the physiological space of the viewing subject, since it is always itself interpreted through symbolic form: whether it be the mythic symbolic form that might structure a sacred area or the modern form of perspective that structures our expectations of "objective" presentation. "It is essential to ask of artistic periods and regions not only whether they have perspective, but also which perspective they have."[26]

Perspective functions as a "symbolic form" since in modern society we cannot help but to understand art, for example, as involving an epistemological distance of subject and object in a manner that Panofsky would argue is shaped by the Western tradition of perspective. Such a relation, however, is also "ontic-ontological" in that we understand our worlds through something like the structure of care that establishes our stakes in the world on the basis of the dichotomy of subject and object created through this particular form. Our sense of possibilities and our relation to the past is palpably shaped by this form. Whereas Heidegger argued that "all interpretations are violent" [*muss jede Interpretation notwendig Gewalt brauchen*], Panofsky understood art historical method to be a means of tempering this violence by situating it in the contemporary contexts that define any being.[27] The art work is one of the few places, as Panofsky's earlier symbolic forms piece had demonstrated and as Cassirer intimated at Davos, where the totality of these contexts overlap and reveal a world on the level of expression, representation, and signification. For both Cassirer and Panofsky, the confrontation at Davos enfigured a new philosophical basis for the methodology of history.

"Symbolic Prägnanz" as the Schema for the Factum of Art as an Example of Ontic-Ontological Meaning

Panofsky's model in "Perspective as Symbolic Form" explains in part why Cassirer cited art against Heidegger as an illustrative sphere of application for his methodology.

A more philosophical rejoinder to Heidegger's claims to "pure ontology," however, is found in the central theoretical development of the third volume of Cassirer's *opus magnum* on symbolic forms, the idea of "symbolic *Prägnanz*." Symbolic *Prägnanz* describes how "perception itself [. . .] by virtue of its own immanent organization, takes on a kind of spiritual [*geistig*] articulation—which, being ordered itself, also belongs to a determinate order of meaning. In its full actuality, its living totality, it is at the same time life 'in' meaning."[28] This conception underlies the use of symbolic forms that Panofsky developed. The German term "*Prägnanz*," which I have kept in the original so as not to have it confused with the substantial associations of the English cognate, has no exact equivalent in English. It combines, as John Krois notes, the German *prägen*—to impress, mint, or give contour—with the Latin *praegnens*, which means full of or ready to give birth.[29] The combination of active and potential energy contained in both terms already gives some sense of Cassirer's meaning. He defines "symbol" as an action that exists in a field of potentiality, not as a thing-like concept with a set meaning.

Although symbolic *Prägnanz* is a concept that might be understood from several vantage points, Cassirer's exposition in his chapter on symbolic *Prägnanz* begins with a critique of Kant's fundamental insight of the necessary synthetic unity of apperception, as well as Husserl's phenomenology, a focus that forms a bridge to the Davos debate discussion of the same theme.[30] The transcendental unity of apperception guarantees the connection of the events of the world—including any mode of self-perception—into a whole and thus enables the fundamental possibility of perception. Where Heidegger will claim the underlying structure of the Kantian schematism can be generalized as a structure of care understood existentially, Cassirer argued that such a structure always irrevocably depends on its particular contents. Judgment and perception are continually staking out a limitation of the plenum of experience under the form of judgment described as "limitative judgment," which Cassirer earlier in the debate calls an "immanent infinity" [*immanente Unendlichkeit*].[31] In Cassirer's view, this form of judgment—as opposed to the empirical sense of objects simply being given or the idealistic stance of ideas being given—was at the heart of all experience. "The fulfillment of finitude," Cassirer states at Davos, "exactly constitutes infinitude. Goethe: If you want to step into infinitude, just go in all directions into the finite." As finitude is fulfilled, that is, as it goes in all directions, it steps out into infinitude. The place of this in Cassirer's work is epitomized by the relation of the particular object or moment of experience in relation to its context. It is this relation that Cassirer argues is epitomized by art, but the relation in general was of principal importance due to its political importance.

Cassirer's definition of symbolic *Prägnanz*'s inflection of "reality" did not end at the modes of expression in myth, aesthetics, and science that he uses in the *Philosophy of Symbolic Forms*, but included political events as well. As Cassirer suggests in a speech celebrating the anniversary of the constitution of Weimar in 1929, political events like the French Revolution are necessarily viewed by a process similar to symbolic *Prägnanz*. In the speech Cassirer contrasts Goethe and Kant's perception of the French Revolution, and the concept of history and of the individual involved in these perceptions. These contrasting views suggest that there are no "empirical facts" outside of "symbolic observation."[32] To an audience violently divided between believers in the Weimar constitution's definition of the individual and state, and those violently against it, the speech must have had a powerful resonance. The implication was that symbolic processes, and their distillation into an immediate mode of perception, pervasively determined what political actors considered "real" in the world around them.

This theme for Cassirer had direct relevance for the growing tide of anti-Semitism and racism in Germany and Europe, for here the problem was that not only assumptions about empirical facts but also *ontological* realities of the more mundane sort were postulated on the basis of symbolic observation. This extension of Cassirer's argument finds further support in the very earliest gestation of the problems dealt with in the philosophy of symbolic forms. In particular, the development of this philosophy might be seen in relation to the so-called Bauch affair that centered around attacks on his teacher Hermann Cohen. At the height of the war hysteria in 1916, Bruno Bauch wrote an article in *Kantstudien* defining German nationality as a matter of birth, and claiming only Germans could understand German philosophy.[33] In a letter to a separate populist/nationalist or *Volkish* journal, *The Panther*, Bauch spelled out more bluntly what he meant specifically: Jewish–German philosophers, including Hermann Cohen, could never properly understand German philosophy since they were not part of the German nation.[34]

Cassirer was furious at this attack, and wrote several drafts of a long response to it—a response that he in the end decided not to publish. Cassirer's answer to Bauch emphasized Bauch's use of visual typologies of nations as the most complete "naturalization" of the concept of culture. Starting out from stereotyped views of Cassirer as "humanist" or proponent of "Enlightenment," one might imagine that Cassirer's argument would criticize Bauch's lack of philosophical acumen and his crude empiricism in judging people by their heritage or how they look, and that Cassirer himself would adopt an idealistic stance that we should judge all people as being inherently equal. The direction of his argument is, however, in nearly the opposite direction. What is marked about Bauch's racism for Cassirer is its formalism and idealism: it first assumes an "essence" to people, around which it then constructs its visual and material object.[35] In this sense Bauch is making a turn toward "materialism," but for Cassirer all materialism is in fact a form of idealism: it idealizes the abstract notion of a single "matter" or of a set of causal material relations underlying experience. Cassirer argued that Bauch has created ideal categories of "German" and German," "non-German," around the theme that "the Jew is a guest in the German house," and that this ideal categorization then determines from the outset how he would see all of the facts about the appearance of German Jews. "Since Jew and German, really or hypothetically, are taken in their 'essence' [*Wesen*] as being fully different, for this reason in the end all of the inner, spiritual/cultural [*geistigen*] *relations* between the two are concealed and explained as mere appearance."[36] The relation of substance and function that Cassirer earlier took up within natural science here has already the distinct political valence that later thinkers would find it held for social science.[37] The "subjects" of Jew and German to which Bauch refers, never actually existed: "they are ghosts."[38] Due to the close relation of ideas and intuition, however, the two form a vicious circle, so that unless carefully examined, intuition will tend to confirm ideas. For this reason, the first demand Cassirer made upon Bauch is that he return to a proper study of history, which would reveal the abstraction of his categories in all of its emptiness. History has to proceed not on the basis of vague philosophical or typological generalizations, Cassirer wrote, but through a study of the "force and details of an individual historical situation and its particularities with which historians brings themselves and others to a living concrete intuition of a situation."[39]

As Cassirer was well aware, such an argument does not, however, really touch on the basic phenomenon raised by Bauch's article or the increasing anti-Semitism that it evidenced. The "objects" created in the forms of nationalism evinced by Bauch are not physical or ideal objects, but affective objects of love or hate. Cassirer's response is suggestively close to the topic that formed the key to his symbolic forms project, which he describes as developing in the winter of 1917, at around the same the time as he was writing his responses to Bauch. Cassirer's theory of symbolic forms centered on his treatment of expression and myth, a theme he was to politicize more extensively in exile in his study of European and Nazi racist ideology, the very partial introduction to which was published as *The Myth of the State*. The key to "expression" is the physiognomic experience as epitomized by myth, which now becomes the most important aspect of human experience, and the most basic way the animate world is revealed to us. The "essence" of a given object is not generated by an ideal category, as Cassirer initially describes Bauch as having done, but as a mythical category.

Cassirer's thought on myth arose in dialogue with the phenomenologist Max Scheler, to whom Heidegger dedicated his Kant book and whose work Cassirer uses to epitomize the general tenets of life-philosophy within which he placed Heidegger's text at Davos. Notwithstanding Cassirer's rejection of Scheler's philosophy overall, Scheler's theory of perception played the key role in Cassirer's description of expression in myth.[40] The passages from Scheler that Cassirer cites in his analysis are in the final pages of a work that was originally published in 1913 with the gripping title *Concerning the Phenomenology and Theory of the Feeling of Sympathy and of Love and Hate*.[41] It is a work that we can readily imagine Cassirer consulting in his attempt a few years later to understand Bauch's creation of the ideal object of the Jew and the German, and how these objects could become linked with affect. This likely early connection would then also prepare the way for its centrality in Cassirer's later rendition in his theory.

In the symbolic forms project, Cassirer describes Scheler's work as the basis of his own theory of the primacy of expression. Cassirer directly quotes Scheler: "It is fundamentally impossible to dissect the unity of an expressive phenomenon (e.g., a smile, a menacing or kindly or affectionate look) into a sum, however great, of the phenomenon in which we perceive the body, or an impression from the physical environment."[42] In this idea, Cassirer finds "confirmation of our fundamental view that what we call reality can never be determined from the standpoint of material alone but that into every mode of positing reality there enters a definite motif of symbolic formation."[43] Cassirer's critique of Bauch had featured the notion of how an "ideal" of the essence of another social group constructs the social intuition of them by other people. Scheler, in turn, directly offers a rereading of the problem of "racial hatred" in a discomforting footnote at the very end of his book. In seeing other people, Scheler says, we are not seeing objects, but grasping an intentional "whole"

[*Ganzheitstruktur*] that already unconsciously imputes a "being" or life to the object. He then continues in a footnote on the nature of racial hatred:

> Empiricism can be very naïve. Hume puzzled over the fact that humans fight and hate, solely because they look "yellow," "black," or "white." This is how he grasped racial fighting and racial hatred! We on the other hand would allow ourselves to take it that Americans don't hate Negroes because they are "black"—since it hasn't been shown that they have any particular hatred towards black handkerchiefs or clothes—but that they sense [*wittern*] a Negro within black skin.[44]

What Scheler here phrases with a worrisome neutrality is precisely the problem that appeared at the forefront of Cassirer's reading of Bauch: how an "ideal" of the essence of another social group constructs the social intuition of them by other people not on the level of thought, but feeling. Whereas for Scheler this immediate "sense" of other people is just taken as a given, or a matter of "instinct," Cassirer's argument demonstrates how this mode of "sensing" a truth is always shaped by form.[45] Cassirer's later *The Myth of the State* would develop this idea, since Cassirer believed that Nazism was distinctive precisely in developing a new "technology" of ritual, action, and images through which it could create the form of perception and hatred that Scheler had described as an autonomous act of the individual.

The creation of "reality" indeed at a level assumed to define all of being and, in a crude fashion, ontology, was thus closely based on an ontic technology and science. Cassirer's position at Davos, as particularly reflected in his later work, anticipated these most practical and gruesome applications of the problems involved. Although the Enlightenment goal of absolute knowledge or truth is lost in Cassirer's philosophy, what is gained is an ability to critique the projected "ontological" reality of the world— whether philosophical or political, precisely in the spirit of Enlightenment and against its Weimar critics. In the immediacy of our experience of art, Cassirer and Panofsky both found the same immediacy open to critique, the same sense of ontological being capable of being understood as the horizon of a particular ontic set of functional relations. For Cassirer and the critical methodology of the study of society, the debate at Davos suggested a level of criticism as relevant to the most awe-inspiring reaches of the new ontology as it was pertinent to the most tangible aspects of political life.

Notes

1. Ernst Cassirer, *Axel Hägerström. Eine Studie zur schwedischen Philosophie der Gegenwart.* (Göteborg: Wettergren & Kerbers Förlag, 1939), 1.
2. Ibid., 2.
3. Karl-Norbert Ihmig, *Cassirers Invariantentheorie der Erfahrung und seine Rezeption des "Erlanger Programms."* (Hamburg: Felix Meiner, 1997) (hereafter cited as *Cassirers Invariantentheorie der Erfahrung*). Gregory Moynahan, "Hermann Cohen's *Das Prinzip der Infinitesimalmethode*, Ernst Cassirer, and Politics of Science in Wilhelmine Germany," *Perspectives on Science* 11, 1 (Winter 2003) (hereafter cited as *Hermann Cohen's Das Prinzip der Infinitesimalmethode*).
4. Jürgen Habermas, "Die befreiende Kräfte der symbolische Formgebung: Ernst Cassirers humanistisches Erbe und die Bibliothek Warburg," in *Ernst Cassirers Werk und Wirkung: Kultur und Philosophie*, ed. D. Frede and R. Schmücker (Darmstadt: Wissenschaftliche Buchgesellschaft, 1996), 99.

THE POLITICS OF HISTORY / 123

Ernst Cassirer, *Substance and Function & Einstein's Theory of Relativity*, trans. William Curtis Swabey, Marie Collins Swabey (New York: Dover Publications, 1953), 237–346. (hereafter cited as *Substance and Function.*); Ihmig, *Cassirers Invariantentheorie der Erfahrung und seine Rezeption des "Erlanger Programms"*; John Michael Krois, "Cassirer, Neo-Kantianism and Metaphysics," *Revue de Metaphysique et de Morale* 97, 4 (Octobre–Décembre 1998): 437–453; John Michael Krois, "Semiotische Transformation der Philosophie: Verkörperung und Pluralismus bei Cassirer und Peirce," *Dialektik* 1 (1995): 61–72; Moynahan, *Hermann Cohen's Das Prinzip der Infinitesimalmethode.*

6. Ernst Cassirer, *Freiheit und Form: Studien zur deutschen Geistesgeschichte* (Darmstadt: Wissenschaftliche Buchgesellschaft, 1991).

7. Martin Heidegger, *Kant and the Problem of Metaphysics*, trans. Richard Taft, ed. J. Sallis, fifth edn. (Bloomington: Indiana University Press, 1997) (hereafter cited as *Kant and the Problem of Metaphysics*); Martin Heidegger, *History of the Concept of Time: Prolegomena*, trans. Theodore Kisiel, ed. J. M. Edie (Bloomington: Indiana University Press, 1992), 213.

8. Martin Heidegger, *Being and Time: A Translation of Sein und Zeit*, trans. Joan Stambaugh, ed. D. J. Schmidt (Albany: State University of New York Press, 1996), 8 (hereafter cited as *Being and Time*).

9. Ibid., 8.

10. Ibid., 48.

11. Ibid., 14.

12. Martin Heidegger, *Kant and the Problem of Metaphysics*, 206.

13. Ernst Cassirer, "Le langage et la construction du monde des objets," *Journal de Psychologie normale et pathologique* 30 (1933): 18–44; Cassirer, 1995.

14. Ernst Cassirer, "Reflections on the Concept of Group and the Theory of Perception," in *Symbol, Myth, and Culture*, ed. D. P. Verene (New Haven: Yale University Press 1979), 136.

15. Thomas Kuhn, *The Structure of Scientific Revolutions* (Chicago: University of Chicago Press, 1996).

16. Cassirer, *Substance and Function*, 266.

17. Ernst Cassirer, *Zur Metaphysik der Symbolischen Formen*, ed. by J. M. Krois and O. Schwemmer. vol. 1, *Ernst Cassirer: Nachgelassene Manuskript und Texte* (Hamburg: Felix Meiner, 1996), 200–201 (hereafter cited as *Zur Metaphysik der Symbolischen Formen*).

18. Heidegger, *Being and Time*, 178.

19. Cassirer, *Zur Metaphysik der Symbolischen Formen*, 205.

20. Ibid., 203.

21. Ibid., 201.

22. Ibid., Michael Friedman, *A Parting of the Ways: Carnap, Cassirer, and Heidegger* (Peru, Ill.: Open Court Publishing, 2000).

23. Heidegger, *Kant and the Problem of Metaphysics*, 206.

24. Cassirer, *Substance and Function*, 309.

25. Erwin Panofsky, "Die Perspektive als 'symbolische Form,' " *Aufsätze zu Grundfragen der Kunstwissenschaft*, ed. H. Oberer and E. Verheyen (Berlin: Wissenschaftsverlag Volker Spiess, 1992), 30 (hereafter cited as *Die Perspektive als "symbolische Form"*).

26. Erwin Panofsky, *Perspective as Symbolic Form*, trans. Christopher S. Wood (New York: Zone Books, 1997), 41.

27. Erwin Panofsky, *Die Perspektive als "symbolische Form,"* 92; David Summers, "Meaning in the Visual Arts as a Humanistic Discipline," in *Meaning in the Visual Arts: Views from the Outside. A Centennial Commemoration of Erwin Panofsky (1892–1968)* (Princeton: Institute for Advanced Study, 1995), 12.

28. ENRfuCassirer, *The Phenomenology of Knowledge*, 202.

29. John Michael Krois, "Cassirer, Neo-Kantianism and Metaphysics," *Revue de Metaphysique et de Morale* 97, 4 (Octobre–Décembre 1998): 437–453, 453.

30. Cassirer, *The Phenomenology of Knowledge*, 193.
31. Heidegger, *Kant and the Problem of Metaphysics*, 201; Martin Heidegger, "Zur Geschichte des philosophischen Lehrstuhles seit 1866," in *Kant und das Problem der Metaphysik*, ed. F. W. v. Herrmann (Frankfurt am Main: Vittorio Klostermann, 1991), 286; Harry A. Wolfson, "Infinite and Privative Judgements in Aristotle, Averroës, and Kant," *Philosophy and Phenomenological Research* 8, 2 (1947): 173–187.
32. Ernst Cassirer, Die Idee der Republikanischen Verfassung: Rede zur Verfassungsfeier am 11. August 1928. (Hamburg: Friederichsen, De Gruyter and Co, 1929), 24–25.
33. Bruno Bauch, "Vom Begriff der Nation. Ein Kapitel zur Geschichtsphilosophie," *Kant-Studien. Philosophische Zeitschrift* 21, 2–3 (1916): 139–162.
34. Bruno Bauch, "Brief, Der Panther," *Deutsche Monatsschrift für Politik und Volkstum* 4 (June 6, 1916): 742–746.
35. Ernst Cassirer Nachlass, Beinecke Library, Yale University, Gen. Ms. 98, Box 52, Folder 1057.
36. Ernst Cassirer Nachlass, Beinecke Library, Yale University, Gen. Ms. 98, Box 52, Folder 1057, 22, see also 19.
37. Pierre Bourdieu, *The Field of Cultural Production: Essays on Art and Literature*, ed. by Randal Johnson. (New York: Columbia University Press, 1993), 4.
38. Ernst Cassirer Nachlass, Beinecke Library, Yale University, Gen. Ms. 98, Box 52, Folder 1057, 19.
39. Ernst Cassirer Nachlass, Beinecke Library, Yale University, Gen. Ms. 98, Box 52, Folder 1057, 22.
40. Cassirer, *The Phenomenology of Knowledge*, 87.
41. Max Scheler, *Wesen und Formen der Sympathie*, ed. M. S. Frings (Bonn: Bouvier Verlag, 1985), 9 (hereafter cited as *Wesen und Formen der Sympathie*).
42. Cassirer, *The Phenomenology of Knowledge*, Scheler, *Wesen und Formen der Sympathie*, 304.
43. Ibid., 87.
44. Scheler, *Wesen und Formen der Sympathie*, 257–258.
45. Ernst Cassirer, " 'Spirit' and 'Life' in Contemporary Philosophy," in *The Philosophy of Ernst Cassirer*, ed. P. A. Schlipp (Evanston/Ill.: The Library of Living Philosophers, 1949), 875.

CHAPTER NINE

PAUL OSKAR KRISTELLER, ERNST CASSIRER, AND THE "HUMANISTIC TURN" IN THE AMERICAN EMIGRATION*

Kay Schiller

Sinnkrise

History bears the marks of the life of those who write it. This truism also applies to the scholarship of the historian of Renaissance philosophy Paul Oskar Kristeller (1905–1999). Moreover, in his scholarly works Kristeller responded, albeit indirectly, to what since Nietzsche became a basic ingredient of the *Weltanschauung* and the academic discourses of the German educated middle class: the perception of a *Sinnkrise*. By this I mean the widespread apprehension of the crisis of the self, meaning, and culture. While the notion of an all-pervasive crisis resulted in the first instance from Germany's rapid industrialization and the experience of World War I and their corollaries, modern technology, mass society and social leveling, the history of the 1930s and 1940s could not but exacerbate it for émigré humanists like Kristeller, not least because they were victimized by a movement that enlisted many of their erstwhile colleagues and almost all of their own students, who convicted them of guilt for the crisis, and who triumphantly proclaimed that their expulsion marked the end of the crisis.

There can be no doubt that scholarship in the human sciences is inextricably linked to the existential preconditions of the scholar at work. Thus it contains elements of self-reflection of the scholar, or, to put it in hermeneutic terminology, each scholarly "expression of life" (Dilthey) also encompasses the autobiography of the writer. References to a level of meaning beyond the topic at hand and toward the life of the scholar are also evident in Kristeller's work. This is of particular relevance in Kristeller's case, because his career after his departure from Germany was advanced not only by Martin Heidegger but also by Giovanni Gentile, giving him sponsorship by the two most respected minds among the supporters of fascist regimes. Although the search for elective affinities in a political sense between Kristeller and the former is fruitless, these special circumstances, and what he made of them, will require further attention.

For the moment, however, let me refer to two details, which indicate the complex weave of emigrant life. After "1968" Kristeller would no longer accept invitations for lectures in his native Berlin. This was not because of the exaggerated violence with which the forces of order had reacted to the West German student movement since the summer of 1967, but because he "fundamentally disapproved" of the protesters' demands for greater participation and the abuse of the academy for political purposes, both at Columbia University, where he was then teaching, and across the Atlantic in Europe.[1] On the other hand, a few years earlier, Kristeller had become a close intellectual friend of Siegfried Kracauer, a writer whose work on film and popular culture had earlier brought him far closer to Max Horkheimer's and Theodor W. Adorno's Institute for Social Research than to the classical tradition. It was Kristeller rather than Leo Löwenthal or any of the other members of that group remaining in the United States who completed, edited and saw through publication Kracauer's last, expressly autobiographical book, *History. The Last Things before the Last.*[2]

For Kristeller, as for Kracauer and many other émigré humanists, the textual space of past ages was not only an object of scholarly inquiry but also source of consolation for the drama of the present.[3] In his learned narratives one detects clues of his identification with one philosophical tradition from antiquity, that is, Platonism, and its Renaissance protagonists. As I intend to show, Platonism was a *philosophia perennis* for Kristeller, that is, the revelation of an immutable and enduring, and, one might add, comforting truth, independent of the vagaries of history. This was because its rational metaphysics provided a link between classical and modern philosophy, between the pre-Socratics and Plato on the one hand, and Kant and Hegel on the other. Although during his academic career Kristeller mostly abstained from disclosing this fundamental belief at the heart of his scholarship, in 1987, more than a decade after he had retired, he admitted that "this tradition has been a rock of intellectual and moral support, much stronger than the numerous fashionable theories and ideologies that have come and gone in rapid succession over the years."[4]

In emphasizing the positive legacy of Platonism, Kristeller's scholarship was part of a wider, internally contested "humanistic turn" in German thought and letters, which emerged since the 1930s. The rediscovery of the "horizons of humanism" in the textual space of European history was a countermove against the figure of the "cold persona," as developed in anthropological, ethical, and aesthetic discourses of the Weimar Republic in the 1920s.[5] While Kristeller was not attracted to this mode of modern thinking, he was, like many of his generation, drawn to Martin Heidegger's "philosophy of existence," itself a reaction to the contemporary experiments in distantiation. While esteem for Heidegger's early thought, in particular for *Being and Time*, remained a constant throughout his life, his intellectual allegiance shifted during the years of emigration to the humanism of Heidegger's great philosophical antipode in the 1920s, Ernst Cassirer, the main representative of the neo-Kantianism of Hermann Cohen's Marburg school and, by then, a German–Jewish émigré himself.

With his *Philosophy of Symbolic Forms* (1923–1929), Cassirer, one of the last representatives of the liberal tradition of German–Jewish intellectuals who drew their inspiration from the German Enlightenment, and who attempted to provide a critical cultural philosophy, meant to address and overcome the all-encompassing

Sinnkrise. Rooted in the Kantian ideals of rationality and cosmopolitan humanity, Cassirer discarded both the pessimistic anthropology of existentialism, as exemplified by Heidegger's "Being-in-the-world" as "Being-toward-death," and the impositions of Darwinist determinism and that of other extractions. As opposed to these, Cassirer's cultural philosophy was grounded in an anthropology of human freedom. He defined man "in terms of human culture" and pointed to man as "animal symbolicum," that is, to man's unique competence to experience the world mediated by symbolic forms like myth, religion, language, art, history, and science. Since for Cassirer the symbolic forms were the manner in which man, a finite being, participated in the infinite, they opened a door toward the liberation of the individual from immediacy and anxiety. To quote from his 1944 *An Essay on Man*, in which he introduced his anthropology and cultural philosophy to an Anglo-American public: "It is symbolic thought which overcomes the natural inertia of man and endows him with a new ability, the ability constantly to reshape his human universe."[6] Cassirer's serene optimism, it appears, was more congenial in aiding a Jewish émigré philosopher from Nazi Germany to cope with his predicament than Heidegger's philosophy of *Endlichkeit.*

A Life of Learning

Born on 25 May 1905, Kristeller was the proverbial "German of Jewish origin," itself a symbolic form of great historical significance for the history of *Bildung.* His family belonged to the well-to-do German–Jewish assimilated bourgeoisie of Berlin. He was brought up by his mother, Alice Magnus, the daughter of a wealthy banker from an old Prussian Jewish family, and his stepfather, the paper manufacturer Heinrich Kristeller, the only father he knew and whose name he assumed in 1919.[7] Deported from Berlin after 1941 on one of the *Alterstransporte,* both of his parents were to die in Theresienstadt.

In 1923 Kristeller followed the neo-Kantian philosopher Ernst Hoffmann, his teacher of classical Greek, who had been called to a chair in Greek philosophy at Heidelberg, to study philosophy, with a specific focus on its history, as well as medieval history, mathematics, and art history. Among his academic teachers at Heidelberg were the philosophers Karl Jaspers and Heinrich Rickert and the medievalists Karl Hampe and Friedrich Baethgen. He also spent two semesters at the university in his native Berlin and went for a semester to Freiburg to hear Husserl, as well as to Marburg to hear Heidegger. He seriously considered completing his degree with a Ph.D. under Heidegger, but eventually settled for a dissertation supervised by Hoffmann. In 1928 he graduated from Heidelberg with a thesis on the founder of neo-Platonism, the Greek philosopher Plotinus.[8]

Because not directly in Heidegger's orbit, Kristeller sidestepped the dilemma that other German–Jewish émigré students of Heidegger like, for example, Hannah Arendt and Karl Löwith faced, that is, to reconcile their profound admiration for Germany's "greatest philosopher" with his zealous engagement for Nazism as rector of Freiburg University in the early 1930s.[9] This is not to say that Kristeller did not fall under the spell of the "Messkirch magician" (Löwith) at all. Evidence for his fascination with Heidegger is not only that his doctoral dissertation was "an existentialist interpretation" of Plotinus,[10] but also that despite his parents' death at the hands of Nazis, he eventually revisited the philosopher and his wife Elfride in Germany in 1973. It can be safely assumed that on that occasion he did not demand an apology, let alone an explanation, for Heidegger's involvement with the regime.[11] Kristeller's rationale for reestablishing friendly relations with Heidegger was that the latter had behaved "decently" [anständig] toward him after 1933, by, for instance, providing him with letters of recommendation and thus facilitating his academic career outside Germany.[12] This was certainly also the reason, why even during the early years of emigration he continued to thank Heidegger in the acknowledgments to his books for advice and help. At the same time, he was quite clear in his correspondence that the infamous rectoral speech of May 27, 1933 was "impossible" [unmöglich], while everything Heidegger wrote afterward, including the Letter on Humanism, "seemed wrong and confused and also contradicting his own earlier philosophy."[13]

Kristeller belonged to the generation of Germans born between 1900 and 1910, which was marked by its generally low chances on the oversubscribed German academic market of the mid 1920s.[14] Classed as a Jew, he also had to cope with anti-Semitism in the ministries and universities. This had negative consequences for Kristeller even before the National Socialists came to power, although the crucial decision was not made by an anti-Semite. Ernst Hoffmann, who would himself be forced into early retirement by the Law for the Restoration of the Professional Civil Service of April 7, 1933 as a Jew, refused to supervise his Habilitation, because he already had one Jewish student, Raymond Klibansky, under his sponsorship and he was convinced that the Heidelberg philosophical faculty would not accept a second one.[15]

With his hopes for an academic future in Heidelberg disappointed, Kristeller returned to Berlin for a degree in classical philology. He studied with Werner Jaeger and Eduard Norden, among others, at Friedrich-Wilhelms-Universität until 1931 when he passed the Prussian state examination to qualify as Gymnasium teacher.[16] With this by-way to an academic career in mind, Kristeller began to work on a Habilitationsschrift in the summer of 1931 on the leading figure of the Florentine Platonic academy, Marsilio Ficino. The project was intended for the Freiburg philosophical faculty, but effectively relied on the personal sponsorship of Heidegger. For obvious reasons, it became impossible to conclude in Germany after 1933.

The story of Kristeller's emigration is quickly told, which is itself an unusual circumstance. Armed with letters of recommendation from Heidegger, Cassirer, whom he had come to know through Hoffmann, and other eminent scholars, he first emigrated to Italy. From early 1934 he lived in Rome, conducting extensive manuscript research at the Vatican and other Roman libraries and scraping through financially as technical assistant to Giovanni Gentile, to whom Heidegger had introduced him. Under his sponsorship, Kristeller first became lecturer in German in Florence and then at Gentile's own University of Pisa and the Scuola Normale Superiore in Pisa.[17]

Gentile, Minister for Public Instruction in the first Mussolini government and "philosopher of fascism," was certainly instrumental in advancing Kristeller's career in Italian emigration, employing him also as codirector of the series of unpublished or rare humanistic texts with the Florentine publisher Leo S. Olschki, where Kristeller's own two-volume edition of Ficino manuscript came out in 1937.[18] But when the adoption of racial decrees by the Italian government cost Kristeller his post in September 1938, not even this powerful member of the fascist establishment could protect him. Although Gentile was unsuccessful in his intervention with Mussolini to have an exception made for "this poor devil" (*questo povero diavolo*), he succeeded in organizing a significant sum to help him cope financially after the loss of his position.[19] As with his most important German mentor, Kristeller was once again forced to distinguish between decent human behavior toward him and the philosopher's public engagement for the cause of his mortal enemies.[20]

In the autumn of 1938, Gentile wrote to American academics in their fields in order to find him a position in the United States.[21] Among the possible employers was the University of Chicago's Classics Department, where Werner Jaeger, who had left Germany because of his Jewish wife, and the Latinist, Berthold L. Ullman, tried to secure him a job and a fellowship from the Oberlaender trust.[22] More promising though were both Kristeller's and his Italian mentors' contacts to the Yale faculty. Among those advocating his cause there were Hermann J. Weigand of the German Department, the émigré historians Theodor E. Mommsen and Hajo Holborn, and, most importantly in terms of academic power, the church historian and Renaissance specialist Roland H. Bainton.[23]

As early as December 1938 Yale's Department of Philosophy extended an invitation to Kristeller to join the faculty for a semester and teach a seminar on Plotinus. Because the American consulate in Naples needlessly delayed the issuing of a non-quota visa for Kristeller for several months, the beginning of his American career was postponed until the spring of 1939.[24] When his contract at Yale expired, he secured a temporary post in the Philosophy Department of Columbia University, where he gradually established himself. Although advancement through the ranks was initially slow for him, in 1948 he finally received tenure. In 1956 he was made full professor and in 1967 he received an endowed chair. He retired in 1972 but continued with his scholarly work until his death in June 1999.

Despite the danger of underestimating the difficulties Kristeller encountered during those years, it is fair to say that compared to other German–Jewish émigré scholars he had a relatively smooth transition from Europe to the United States. This was certainly due to the fact that he was thoroughly trained, exceptionally gifted, and well recommended by prominent non-Jews in both philosophy and classical philology. These two academic disciplines, in combination, possessed a special aura of legitimacy in Germany and Italy, as well as in the United States, with German university credentials in these fields having a unique value to the generation of academics in positions of power. In any event, his American career was a "success story" in terms of both scholarly creativity and recognition. He leaves behind a large oeuvre as a historian of Renaissance (and classical) philosophy, as an editor of Renaissance philosophical texts, translations and commentaries, and, most importantly for future generations of scholars, as an author and compiler of the *Iter Italicum*, a monumental finding aid for Italian Renaissance manuscripts in European and American archives and libraries.[25] In the latter years of his career Kristeller was showered with academic honors both in the United States and Europe, including Germany. He was presented seven homage volumes and received no less than ten honorary doctorates, as well as a number of medals and prizes from scholarly academies and learned societies in different countries. In an obituary, a prominent colleague concluded: "He may prove to have been, after Jacob Burckhardt, the most important student of the Renaissance in modern times."[26]

Ficino

After his arrival at Columbia, Kristeller was at first predominantly concerned with the continuation of his studies on the Platonism of the Italian Renaissance, beginning with an English translation of his monograph on Ficino, which he had completed in Italy in 1938. The book, entitled *The Philosophy of Marsilio Ficino*, came out in 1943. It was a historical analysis of the entire system of Ficino's philosophy, his metaphysics, psychology, and philosophy of religion.[27]

Until this day Marsilio Ficino is best known for two accomplishments: in the first instance, for the pivotal role he played in the foundation of the Platonic Academy in Florence. Based on the original academy in Athens of some 1,200 years earlier, this was an informal circle of Ficino's friends closely linked to the Medici court, in which Plato's philosophy was discussed and through which it was spread among the contemporaries.[28] Second, Ficino is still recognized, because he introduced the love theory of Plato's *Symposium* and *Phaedrus* to the Renaissance. While inner experience or contemplation was the central concept of Ficino's Platonism, Socratic or Platonic love provided the spiritual bond of friendship among the members of the academy, that is, the fellowship of those who participated in the contemplative life.

For Kristeller, of course, Ficino stood for much more than those two achievements. As he wrote in the introduction to his book:

Ficino's Platonism is not a philosophical conception that just happened to appear during the Renaissance, it is, so to speak, the Renaissance become philosophical—in other words, the philosophical expression and manifestation of its leading idea.[29]

What were Kristeller's reasons for this rejection of the historicizing type of interpretation prominent since the writings of Burckhardt? Generally speaking, he thought it essential to take Ficino's philosophy seriously, on its own terms. This was, first, because at the heart of Ficino's Platonic speculation lay a theory of the immortality of the soul. While the belief in immortality as a religious doctrine belonged to the standard repertoire of all Christian and Platonic thinkers, Ficino's claim that it could be rationally demonstrated was unprecedented. Second, and related to this, Ficino developed a doctrine of human dignity, which in contrast to the medieval emphasis on God, placed man and man's rational soul at the center of the hierarchy of the universe.

To be sure, Ficino had no intention of proclaiming a this-worldly philosophy. The main purpose of his metaphysical speculation was to meet the spiritual needs of those who wanted to reconcile their Christian beliefs and the study of classical antiquity. While he emphasized man's rationality and central role, he also demonstrated that even "though Platonic philosophy ha[d] its own authority and tradition, it [was] in no way opposed to Christian doctrine and tradition."[30] As Cassirer put it in his long and very positive review of Kristeller's book: "Personally Ficino was no 'free thinker.'

He did not defend, he did not even conceive the ideal of the 'autonomy of reason' or of a secular philosophy. He never went beyond the limits of a '*philosophia pia*.' "[31] Yet he advocated religious tolerance. To quote Cassirer again:

> He strove for a universal religion not for a universal church. Everyone who worshiped and loved God was welcome. There were no heretics in this new religion. For what is essential in religious life is not any dogmatic formula. According to Ficino the difference between formulae, between external signs and symbols, does not endanger the unity of faith; on the contrary, it confirms this unity.

In a 1960 conference paper on the Platonic Academy Kristeller echoed Cassirer's assessment by writing that Ficino's was "a doctrine that advocated harmony and tolerance in a period torn by the religious conflicts preceding and following the Reformation."[32] For the Florentine philosopher, contemplation meant "a gradual ascent of the soul towards a highest goal, the direct knowledge of god." Can one not discern in this statement an exile's preference for the life of the mind, an escape from the cataclysms of the present? Yet this invocation of the somewhat soft neo-humanism characteristic of the public face of Cassirer's and Kristeller's American careers does not dig very deep.

For a more conclusive answer, it is worth approaching these questions from a different direction. One can look more broadly at how Kristeller proceeds in his scholarship. In his Ficino book, as well as in his later works, Kristeller was first and foremost a historian of textual and intellectual genealogies. As a philologist he focused on the textual transmission of Ficino manuscripts, while as a historian of ideas his emphasis here was primarily on the influence, which the Platonic tradition exerted on the Florentine philosopher. Like Cassirer in his own forays into the history of ideas and opposed to cultural historians of the Renaissance in the tradition of Burckhardt, Kristeller concentrated on the transmission of philosophical thought in a relatively narrow sense.

This went hand in hand with a relative disregard for the wider political and social context of Medici Florence within which the philosopher and his Platonic Academy were situated. In his assessment of Ficino's "metaphysics of reason," Kristeller did not regard it as a defect that it had an apolitical bent to it, as Ficino "was not interested in political problems."[33] Moreover, it did not concern him that Ficino's metaphysical speculations were only made possible by the patronage of the Medici family and that it fitted well with the interests of their authoritarian political regime to distract the attention of the population from the affairs of their state between the end of the Florentine Republic in 1434 and Savonarola's revolution sixty years later in 1494.

Kristeller's focus on philosophy in general and Ficino's neo-Platonism in particular also led him to play down the importance of "classical humanism," the leading intellectual movement, which, according to Burckhardt, had been instrumental in setting the Renaissance apart from the Middle Ages. While Kristeller acknowledged that humanism was original to this epoch, the humanists for him were mainly representatives of a rhetorical and poetical culture, in short, a nonphilosophical culture. Of course, these were learned men whose efforts revolved around the revitalization of the rediscovered literature and culture of Greece and Rome and the *studia humanitatis*, grammar, rhetoric, history, poetry, and moral philosophy. However, in terms of earnest metaphysical speculation their contribution was rather limited. They were not to be taken seriously, for their works were "amateurish" and not adequately grounded in reason.[34]

Implicit in Kristeller's approach was a twofold challenge: first, to the predominant paradigm of German historical scholarship on the Italian Renaissance, along the lines of Burckhardt, which emphasized the importance of the epoch in terms of universal history as the cradle of the modern spirit where the proverbial "discovery of the world and of man" had occurred. Although Kristeller did not go as far as to dismiss the notion of the Renaissance as a separate epoch altogether, he stressed the importance of continuities from the Middle Ages, for instance with regards to humanist grammar and rhetoric.[35] Incidentally, Kristeller's perspective certainly fitted well with prevailing trends in American academic scholarship without damaging the wider appeals of neo-humanism among the onlookers—and funders—of academic work, substantially aiding his acceptance in U.S. academia.[36]

Second, and crucial in terms of finding an answer for the above question, Kristeller in his scholarly work discarded contemporary attempts to instrumentalize the humanist tradition for the present. One example of this from interwar Germany was the vociferous efforts of Kristeller's former academic teacher, Werner Jaeger, to reactivate Greek antiquity against the noisy political conflict of the Weimar Republic. In his own version of the "humanistic turn," Jaeger, a conservative classicist who saw himself as a semipolitical educator, wanted to imbue German politics, society, and culture with classical values by way of a "third humanism" (following on from Renaissance humanism and German neo-humanism).[37] After 1933, initiatives like this one, meant to overcome the supposed crisis of meaning, were easily adapted to support the Nazis, by seizing on the more ruthless aspects of Jaeger's protagonists, Plato and Thucydides.

For Kristeller, historical anachronisms of this kind were simply unacceptable. What were cultural phenomena of the past could only be understood adequately, if one resisted the "temptation" of emphasizing the transhistorical "human relevance of certain problems."[38] Against self-declared "renovators" of the humanist tradition, he insisted on a Kantian notion of pure knowledge and scholarship, uncontaminated by the concerns of the present. *Wissenschaft* for Kristeller was "the problem of universals, the criterion of truth" and "the range of human knowledge, the plain facts ascertained by experience and reason [which] cannot be contradicted by an appeal to conventional and fashionable opinions."[39] Yet political aspects of scholarly disputes were not ignored by Kristeller. Accordingly, he and his peers derided the "third humanism" exported by their uncomfortable fellow émigré, Werner Jaeger.[40] In 1934 one of his closest friends wrote bitterly to him: "Have you read Jaeger's Paideia yet? There are quite funny NS'isms in it!"[41] Another friend, the émigré art historian Erwin Panofsky, parodied a German nursery rhyme to suggest that behind "third humanism" and *Paideia*, National Socialism could be heard "rustling in the straw."[42]

However, the rejection of anachronistic exploitations of the "classical ideal" did not mean that Kristeller and Panofsky maintained that all achievements of history had to be relegated to a dead past. There were indeed traditions from history worth preserving for the present. It was crucial which part of the heritage was at stake, how and by whom the "rescue effort" was undertaken, and to what end. The preservation of a deserving tradition had to be conducted in a safe and protected space, among a circle of learned friends dedicated to philosophy and removed from the noises and dangers of politics and practical life, an "ivory tower" not unlike the Florentine Academy.

Renaissance Platonism, as Kristeller put it quite lyrically in the 1948 edition, was "a this-worldly religion of the imagination—attractive in contour and wistfully reminiscent of another world, like the Platonism of Botticelli's pencil and, like it also, thin and disembodied and ever trembling on the verge of the Christian mystery."[43] This bold and deeply personal assertion is the earliest sign that permits an extended exploration of a source, which may reveal more about the overlap between autobiography and scholarship in his work—the historical inquiries and commentaries of Ernst Cassirer, with whom he was closely linked during the first decade or so of emigration.

Such overlap inherently creates tensions between the strict asceticism of scholarship that Kristeller invoked against the diffuse and politically dangerous idealizations of the "third humanism" on the one hand, and, on the other, the contribution of scholarship to Kristeller's own search for meaning in such cruel times. The lead must be followed.

Cassirer and Platonism

What then were the links? There was, in the first instance, the personal contact, which Kristeller and Cassirer maintained during the years of emigration from 1933 until Cassirer's death in April 1945. They regularly communicated by way of letters and postcards and met frequently once they both reached the United States. After resigning from his professorship in Hamburg even before the Nazi regime could force him to do so, Cassirer left Germany and spent the next eight years in England and Sweden, first at All Souls College in Oxford and then at Högskala University in Göteborg. He eventually emigrated to the United States in 1941, where he taught at Yale until reaching retirement age three years later. During the academic year 1944–1945 he was a visiting professor in Kristeller's department at Columbia University.

Faced with the problems of life as a refugee himself, Cassirer nevertheless did everything in his power to provide support for younger and less well-known fellow émigré humanists like Kristeller. It was, for example, due to Cassirer's recommendation that Kristeller, although living in Italy, obtained a research grant in 1935 from the London based Academic Assistance Council (renamed to Society for the Protection of Science and Learning in 1936), the main British philanthropic organization in aid of German–Jewish refugee scholars.[44]

Second, there was Kristeller's admiration for Cassirer's exploratory studies in his own chosen period, most prominently *Individual and Cosmos in the Philosophy of the Renaissance*, his 1926 book on Nicolaus Cusanus's thought, which was among the first to acknowledge the existence of a proper "philosophy of the Renaissance."[45] This made a monograph on Ficino by a historian of philosophy both possible and desirable. For Cassirer, in turn, Kristeller's book filled an important gap in the history of philosophy. He praised it as a work that linked Greek antiquity with the German Enlightenment and idealism, Socrates with Kant and Goethe. As Cassirer pointed out in an important 1943 article on Pico della Mirandola, the history he had himself uncovered (and to which Kristeller had contributed his highly original work) stretched from the neo-Platonists thinkers of antiquity, like Plotinus and Porphyrius, to Renaissance Platonists like Cusanus, Ficino, and Pico, to the Cambridge Platonists of the seventeenth century and Shaftesbury, and in Germany via Leibnitz to Kant, Winckelmann, and Goethe.[46]

Ficino's metaphysical speculation, presented comprehensively for the first time in Kristeller's monograph, was particularly relevant, because this effort set the stage for Pico's Oration *De hominis dignitate*, for Cassirer one of the key-texts of the Western tradition. While for Ficino, in truly Platonic fashion, man's excellence still consisted in the role man's rational soul played as the center of a hierarchically structured universe, his friend Pico took this idea one step further by setting man altogether apart from this hierarchy. His concept of human dignity designated an exceptional and privileged position for man, because for Pico man was different from both the natural and the spiritual world. As Cassirer put it:

> This is man's privileged position: unlike any other creature, he owes his moral character to himself. He is what he *makes* of himself—and he derives from himself the pattern he shall follow.[47]

Whereas for Ficino, man's likeness and resemblance of God was still dependent on divine grace, for Pico it is "an achievement for [man] to work out: it is *to be brought about* by man himself."[48] To quote Pico: Man is *sui ipsius* [. . .] *plastes et fictor*. He is, in Cassirer's translation, "the 'sculptor' who must bring forth and in a sense chisel out his own form from the material with which nature has endowed him."[49]

Pico's insistence on human freedom and dignity resonated in Cassirer's own thought, as epitomized in *An Essay on Man*, a pivotal product of his emigration. In addressing the crisis of the modern self—the *Sinnkrise* that had been the younger Kristeller's starting point—Cassirer was engaged in developing an anthropology of freedom. Its aim was to uncover what he considered to be the true meaning of human existence, as against the impositions of *Existenzphilosophie* and determinism prominent during the Weimar years, that is, the liberation of the individual from immediacy and anxiety by way of symbolic thought. In his famous 1929 disputation with Heidegger he emphasized the ultimate duty of philosophy "to allow man to become as free as possible."[50]

If, as Cassirer wrote in *An Essay on Man*, "[h]uman culture taken as a whole [could] be described as the process of man's progressive self-liberation" through symbolic forms like language, art, religion and science,[51] then the Renaissance and its philosophers played a particular role in this process. In his 1926 book on Cusanus he had already interpreted this epoch as crucial for the initiation of modern thought.[52] In Cassirer's view, Renaissance philosophy rediscovered what classical philosophy knew all along, that is, the creative potential of man, man's capacity for symbolic thought.

For Cassirer, then, while the renewal and transformation of Platonism by Cusanus, Ficino and Pico marked the beginning of modern thought, their metaphysical speculation was inseparably linked with the rediscovery of a promise from antiquity. This was the prospect of liberation from the limitations of man's finite existence through man's "power to build up a world of his own, an 'ideal' world."[53] For Cassirer, this legacy from Platonic philosophy was taken up and extended by the Enlightenment and German idealism. And, as his stirring invocation of the themes meant to show, it had lost none of its relevance in the twentieth century.[54]

While Kristeller never spoke as Cassirer did, as such language would have gone against his methodological exclusion of anachronism, I think that I have shown that the underlying conception nevertheless shines through his technically much more demanding writings on the Florentine Platonic Academy. Autobiography, the personal struggle, leaves the marks of its formative effects. Before 1987 Kristeller comes closest to Cassirer's broader rendering of the meaning of the Platonist Renaissance legacy where he admits parenthetically that Platonism for him was indeed a *philosophia perennis*, what Cassirer characterized as "the revelation of an enduring Truth, in its main features immutable [. . .] handed down through the ages, but generated by no age [. . .], because, as something which eternally is, it is beyond time and becoming."[55] In the closing, characteristically self-contained sentence of his 1960 paper on the Platonic Academy, Kristeller writes:

> Finally, if we are inclined to consider the history of Platonism in the West as a kind of *philosophia perennis* (and I must confess that I share this inclination), we shall have to admit that the Florentine Platonism of the Renaissance, with all its defects and weaknesses, represents one of the most important and most interesting phases in the history of this philosophical tradition.[56]

It should have become evident that for Cassirer and Kristeller (and Panofsky for that matter) the Platonism of the Florentine Academy, with the insistence on tolerance, human dignity, and the utopian promise of human freedom that they found in it, proved to be a source of comfort and consolation against the catastrophe that in so many ways determined the course of their lives. This was their version of the "humanistic turn." Whatever may have been the rhetorical strategy of Kristeller's friend Cassirer in making his philosophical life's work sound in a strange land in the hardest of times, Kristeller's pursuit of their shared objectives for the most part required a strict, aristocratic withholding of didactic uplift in his utterance. This self-denial was no less profound a sign of his intense feelings under conditions of exile as were the resonant exhortations of the older, more famous humanist, Cassirer. The thought shared among Kristeller, Panofsky, and Kracauer, as expressed long after Cassirer was dead, was that the distance that constituted an Academy in the modern age of pervasive ideologies and publicity engines could only be sustained by the utmost in disinterested attentive accuracy of a kind unimagined by the Renaissance masters, so profoundly at home in Florence. Among these exiled heirs of Moses Mendelssohn's initiation of what Habermas called the "abysmal yet fertile relationship of the Jews with German philosophy," the philosopher emerged as the stranger.

Kristeller could not have gone further away from the activism of his early mentors, Heidegger and Gentile.[57]

Notes

* I owe many of the ideas for this article to the collaboration with Gerald Hartung on our common project *Weltoffener Humanismus. Philosophie, Philologie und Geschichte in der deutsch-jüdischen Emigration.* Thanks must also go to Oliver Zimmer and Warren Boutcher for their comments, as well as to David Kettler and two anonymous readers for their suggestions.

1. Kay Schiller, Interview with Paul Oskar Kristeller, New York, 4/9/1993. Transcript.
2. See Siegfried Kracauer, *History. The Last Things before the Last* (New York: Oxford University Press, 1969), v–xi.
3. For other examples see Kay Schiller, *Gelehrte Gegenwelten. Über humanistische Leitbilder im 20. Jahrhundert* (Frankfurt: Fischer, 2000) and "Hans Baron's Humanism," *Storia della Storiografia* 34 (1998): 51–99.
4. Paul Oskar Kristeller, *Marsilio Ficino and His Work after Five Hundred Years* (Florence: Leo S. Olschki, 1987), 18.
5. See Helmuth Lethen, *Cool Conduct. The Culture of Distance in Weimar Germany* (Berkeley, Los Angeles: University of California Press, 2002).
6. Ernst Cassirer, *An Essay on Man. An Introduction to a Philosophy of Human Culture* (New Haven, London: Yale University Press, 1944), 62 (hereafter cited as *Essay on Man*). For a recent assessment of this book in relation to the development of Cassirer's thought both before and during his emigration, see Gerald Hartung, "Anthropologische Grundlegung der Kulturphilosophie. Zur Entstehungsgeschichte von Ernst Cassirers *Essay on Man*," *Kulturwissenschaftliche Studien* 6 (2001): 2–15.
7. Paul Oskar Kristeller and Margaret L. King, "Iter Kristellerianum: The European Journey (1905–1939)," *Renaissance Quarterly* 47 (1994): 907–929, 907–910 (hereafter cited as *Iter*).
8. Paul Oskar Kristeller, *Der Begriff der Seele in der Ethik des Plotin* (Tübingen: Mohr, 1930).
9. Richard Wolin, *Heidegger's Children. Hannah Arendt, Karl Löwith, Hans Jonas, and Herbert Marcuse* (Princeton: Princeton University Press, 2001), 9.
10. John Monfasani, "Obituary: Professor Paul Oskar Kristeller," *The Independent*, London, 24/7/1999 (hereafter cited as *Obituary*).
11. Kristeller to M. Heidegger, 9/4/1973, Kristeller Papers (hereafter cited as *KP*), Box 19, Folder Healey-Heitmann, Butler Library, Columbia University, New York.
12. See "Empfehlung für Dr. Kristeller, Freiburg 30. April 1933," in Martin Heidegger, *Reden und andere Zeugnisse eines Lebensweges, 1910–1976* (Frankfurt: Vittorio Klostermann, 2000): 89 and Kristeller and King, *Iter*, 918.
13. Kristeller to R. de Rosa, 19/12/1987 and 12/1/1986, *KP*, B. 12, F. de Rosa, Renato.
14. Detlev J. K. Peukert, *The Weimar Republic. The Crisis of Classical Modernity* (London: Allen Lane, 1991), 87–88.
15. Kristeller and King, *Iter*, 915. On Hoffmann's forced retirement see Dorothee Mussgnug, *Die vertriebenen Heidelberger Dozenten: Zur Geschichte der Ruprecht-Karls-Universität nach 1933* (Heidelberg: Carl Winter, 1988), 68–69.
16. Kristeller and King, *Iter*, 916.
17. On Gentile see Gabriele Turi, *Giovanni Gentile. Una Biografia* (Florence: Giunti, 1995).
18. Paul Oskar Kristeller (ed.), *Supplementum Ficinianum: Marsilii Ficini Florentini Opuscula inedita et dispersa*, 2 vols. (Florence: Leo S. Olschki, 1937).
19. G. Gentile to B. Mussolini, 2/9/1938, *KP*, B. 16, F. Letters to and from G. Gentile concerning P. O. Kristeller.
20. See Turi, "Giovanni Gentile, Oblivion, Remembrance, and Criticism," *Journal of Modern History* 70 (1998): 919–933, 932.

21. See for example D. Cantimori to R. H. Bainton, 25/9 and 17/11/1938, Roland Bainton Papers, Record Group 75, Yale Divinity Library, New Haven, Connecticut and R. H. Bainton to D. Cantimori, 24/10/1938, *KP*, B. 2, F. Letters to and from R. Bainton.
22. W. Jaeger to Kristeller, 13/10/1938, *KP*, B. 23, F. Jaeger, Werner and B. L. Ullman to G. Gentile, 8/9/1938 and to Kristeller, 7/10 and 14/11/1938, *KP*, B. 51, F. Ullman, B. L.
23. Th. E. Mommsen to Kristeller, 11/10/1938, *KP*, B. 36, F. Modoni-Mommsen, H. J. Weigand to Kristeller, 22/9, 30/10, and 12/12/1938, *KP*, B. 53, F. Weigand, Hermann (1) and R. H. Bainton to G. Gentile, 23/12/1938 and 1/1/1939, *KP*, B. 16, F. Letters to and from G. Gentile concerning P. O. Kristeller.
24. R. H. Bainton to Thomas D. Bowman, American Consul at Naples, 6/1/1939, *KP*, B. 2, F. Letters to and from R. Bainton.
25. Paul Oskar Kristeller (ed.), *Iter Italicum: Accedunt Alia Itineraria. A Finding List of Uncatalogued or Incompletely Catalogued Humanistic Manuscripts of the Renaissance in Italian and Other Libraries*, 7 vols. (London, Leiden: The Warburg Institute, E. J. Brill, 1963–1996).
26. Monfasani, *Obituary*.
27. Paul Oskar Kristeller, *The Philosophy of Marsilio Ficino* (New York: Columbia University Press, 1943) (hereafter cited as *Philosophy*).
28. For a thorough historical re-assessment of the academy, its structure, the extent of its membership, its tenuous connection to the Medici court and the limited role of Plato in its intellectual life, see James Hankins, "The Myth of the Platonic Academy of Florence," *Renaissance Quarterly* 44 (1991): 429–475.
29. Kristeller, *Philosophy*, 23.
30. Ibid., 27.
31. Ernst Cassirer, "Ficino's Place in Intellectual History," *Journal of the History of Ideas* 6 (1945): 483–501, 489.
32. Paul Oskar Kristeller, "The Platonic Academy of Florence," *Renaissance News* 14 (1961): 147–159, 148–149 (hereafter cited as *Platonic Academy*).
33. Kristeller, *Philosophy*, 15.
34. Ernst Cassirer, Paul Oskar Kristeller, and John H. Randall, Jr., (eds.), *The Renaissance Philosophy of Man* (Chicago, London: The University of Chicago Press, 1948), 5–8 (hereafter cited as *Renaissance Philosophy*). The edition, which contained philosophical texts by Petrarch, Valla, Ficino, Pico, Pomponazzi, and Vives, was already conceived before Cassirer's death on April 13, 1945. Cassirer was originally meant to write the general introduction to the volume.
35. See Paul Oskar Kristeller, "Humanism and Scholasticism in the Italian Renaissance," *Byzantion* 17 (1944–1945): 346–374.
36. For the "revolt of the medievalists" under the leadership of Charles H. Haskins, the doyen of medieval studies in the United States, against the Burckhardtian image of the Renaissance, see Charles H. Haskins, *The Renaissance of the Twelfth Century* (Cambridge, Mass.: Harvard University Press, 1927). Kristeller's approach also fitted well with the "history of unit-ideas," as advocated by Arthur O. Lovejoy, the editor of the *Journal of the History of Ideas*, founded in 1940. It only took until 1943 before he was invited to join its editorial board.
37. See Suzanne Marchand, *Down from Olympus: Archeology and Philhellenism in Germany, 1750–1970* (Princeton: Princeton University Press, 1996), 319–330 and Beat Näf, *Von Perikles zu Hitler? Die athenische Demokratie und die deutsche Althistorie bis 1945* (Bern, Frankfurt: Peter Lang, 1986), 89–92, 187–191.
38. Paul Oskar Kristeller, "Studies on Renaissance Humanism during the last Twenty Years," *Studies in the Renaissance* 9 (1962): 7–30, 9–10.
39. Paul Oskar Kristeller, "Scholarship, Past and Future." Typescript, Medieval Seminar at Columbia University, 13/4/1976, p. 5, *KP*, B. Columbia University Subject Files 3, F. Columbia University Medieval Seminar.

40. See Werner Jaeger, *Paideia. Die Formung des griechischen Menschen*, 3 vols. (Berlin, Leipzig: de Gruyter, 1934, 1944, 1947); translated into English as *Paideia: the Ideals of Greek Culture*, 3 vols. (Oxford: Blackwell, 1939, 1943, 1944).

41. "Hast Du schon Jaeger's Paideia gelesen? Es sind ganz lustige NS'ismen drin!" E. Abrahamson to Kristeller, 10/4/1934, *KP*, B. 1, F. Abrahamson, Ernst (1).

42. Erwin Panofsky, "Alte Witze." Manuscript, no date, William S. Heckscher Papers, Series II, Folder 21, The Getty Center for the History of Art and the Humanities, Santa Monica.

43. Cassirer, Kristeller, and Randall, *Renaissance Philosophy*, 6–7.

44. Cassirer described his activities on Kristeller's behalf in the United Kingdom in a number of letters and postcards to the latter: Cassirer to Kristeller, 21/9/1933, 12/1, 22/4, and 4/5/1934, *KP*, B. 8, F. Cassirer, Ernst. See also Cassirer to Fritz Saxl, 13/8/1935, General Correspondence, The Warburg Institute Archives, London.

45. See Ernst Cassirer, *Individuum und Kosmos in der Philosophie der Renaissance* (Leipzig: Teubner, 1926).

46. Ernst Cassirer, "Giovanni Pico della Mirandola. A Study in the History of Ideas," *Journal of the History of Ideas* 3 (1943): 123–144 and 319–346, 345 (hereafter cited as *Giovanni Pico della Mirandola*).

47. Ibid., 320.

48. Ibid., 321.

49. Ibid., 333.

50. "Protokoll der Davoser Disputation zwischen Ernst Cassirer und Martin Heidegger," in Martin Heidegger, *Kant und das Problem der Metaphysik* (Frankfurt: Vittorio Klostermann, fourth edn. 1973), 246–268, 259.

51. Cassirer, *Essay on Man*, 228.

52. Heinz Paetzold, *Ernst Cassirer—Von Marburg nach New York. Eine philosophische Biographie* (Darmstadt: Wissenschaftliche Buchgesellschaft, 1995), 123.

53. Cassirer, *Essay on Man*, 228.

54. See Gerald Hartung, "Einleitung," in Ernst Cassirer (ed.), *Die Philosophie der Aufklärung* (Hamburg: Felix Meiner, Reprint 1998): vii–xxiii, xix.

55. Cassirer, *Giovanni Pico della Mirandola*, 124.

56. Kristeller, *Platonic Academy*, 159.

57. Jürgen Habermas, "The German Idealism of the Jewish Philosophers (1961)," in *Philosophical-Political Profiles* (Cambridge, Mass.: MIT Press, 1983): 21–43, 22.

Chapter Ten

"The Reparation of Dead Souls"—Siegfried Kracauer's Archimedean Exile—The Prophetic Journey from Death to Bildung

Jerry Zaslove

The ancient historians used to preface their histories by a short autobiographical statement—as if they wanted immediately to inform the reader of their location in time and society, that Archimedean point from which they would subsequently set out to roam the past.

—Siegfried Kracauer[1]

Dragged into Exile and Utopian Excavation

Since the discovery of Siegfried Kracauer by the generation that was first spellbound by the writings of Bertolt Brecht, Walter Benjamin, T. W. Adorno, Ernst Bloch, and Hannah Arendt, the Weimar legacy of the exile has been drawn on many times as a paradigm of the intellectual whose life was unfinished because so much of it was left in the past. Those whose bridges to the past were burned by the forgetfulness of the age might include Kracauer. Siegfried Kracauer's work has fascinated for its outsider-ness and the difficulty of placing him in any canonical role as a founder of any move-ment or school. The literature on him has exploded to the point that one must wonder whether his reputation will forever be linked with some other than the exile generation and if so, which one?[2] He seems so young and so old at the same time. Yet he remains a member of the generation of Klaus Mann, Stefan Zweig, Lion Feuchtwanger, whose condition is described by Günther Anders, who wrote

> There was not one among us who did not one day stand still at some corner of some city and discover that the calls and noises of the world suddenly sounded as if they were meant for others—who did not have the experience of no longer being there [. . .]. The exile suicides simply sealed this loss of existence.[3]

Kracauer did not commit suicide, nor did his emigration allow him to avoid the no-mans-land that the exile carried with him in the way that Brecht once said of the exile like himself, "who carried a brick with him to show the world what his house had been like."[4]

Kracauer experienced many exiles in his life, and my approach to his work is meant to establish that everything he wrote was in some way a snapshot of being an exile, or approaching exile, or thinking about the exile condition. The reader will see that I approach his work as if it were a continuity around several "Archimedean points." His writings in America were in a sense parables of a hoped for return—not to Germany but to a radical concept of *Bildung* for the future. In exile he never—perhaps to the dismay of some—wrote about the politics of Jewish identity or Zionism. He did not do sociology. Neither did he do history as such. His work cannot be understood as a meta-existential statement about philosophy or the Jewish–German symbiosis. Yet behind everything that he wrote or thought lies the legacy of the stranger, what Hannah Arendt referred to as the Jew as pariah and parvenu. My approach to Kracauer begins with his characterization of modernity as the creation of alienated, exterritorial individuals and destroyed collectives, and my approach applies to him the same assumptions that he applied to others. I differ however from many other commentators by interpreting his work as if a personal myth, an inscape lines the interior of his work and that this is the optical measure through which one must see his politics of memory.

There are many ways to read Kracauer's story. I prefer to use his own method of seeing a life history in overlapping phases and archimedian points that lever into the biography of a generation. The exilic phases of his life are, first, his cosmopolitan anxiety in his Weimar attempt to apply Kantian principles to social phenomena and inner experience; second, his engaged writing "praxis" for the newspaper and the public sphere in a variety of genres; third, his eviction and exile from Germany, being forced to go to Paris and lose his bearings in an exterritorial state of existence; fourth, his escape to America where he poured the idea of exile into *Caligari* as a study of German film and Fascism as a public sphere impoverished in *Bildung*; and, finally, in the United States, finding the film again as the "Oklahoma nature theatre" of America, where exile was no longer the "ornament of the masses" but became the optical principle of reading and writing history through modern film. In conclusion I write about his use of a colportage structure for memory to the rediscovery of history as the medium in which his struggle for existence could once again recover his balance between the essay form and history and photography.

The interest in Kracauer accompanies writings from the post-exile generation that has turned away from Marxism and has entered the world of cultural criticism, *Erfahrung*. This means in part that we are seeing the displacement of the European exilic consciousness into the notion of the wayward or homeless text that is linked to the idea of the diaspora. At a time in history when ethnicity has become a positive form of recognition, the diasporic consciousness about dispersed peoples and races pays attention to racialized differences. Formerly this was a small stream in intellectual worlds limited to persecution, coercion and exile. Today, with the growth of the global movements and neo-market forces transcending the borders and most forms of cultural production, exile consciousness comes to itself through different mediations. It is problematic whether Kracauer can be used to reinforce a politics of exile.

However, my own view is that there is an anarcho-communitarian stream in his work—perhaps reflected more in Kracauer than in Adorno or Benjamin—toward a post-identity concept of *Bildung*, even as the idea of the "cosmopolitan" may be in disrepute. However, here this essay cannot follow that stream to its source.[5] The attention to Kracauer does not only speak to the academic turn to film, which dominates the intellectual landscape of this generation, but to something in his work that calls to us in the way one returns to a landscape that we return to again and again but has no name—perhaps in Kracauer's own sense we turn to the film as a "ruin of ancient beliefs."[6] Kracauer's way of releasing and liberating the meanings in culture is created by his essayist style. The novel or photography properly are broken texts that rely on a context that is not always obvious to the reader who expects a particular genre. He himself refused to define himself as a sociologist, historian, or novelist, and it was not until last work, *History. The Last Thing Before the Last*, where we can see how and why he made peace with history both as a house of slaughter and a house for the exterritorial orphans who went into exile in search of a second chance.

Only late in his life did he see himself a writer who searched the abyss of his memory for any reopening of the question of whether it was possible to write history after the multiple catastrophes of the loss of friends, position, language, and imagery that made up one's inner world. In keeping with Kracauer's "*Geschichte-wollen*" as *Kunstwollen* his last book on history is a gift to the future generations who might want to find a way to rewrite their future based on how one thinks of exile. He addresses his last work to generations who were forced to leave and those who did not, and, also, to those whom, as Henry Pachter wrote, lived on the dividing line between aliens in exile and those who profited by remaining behind within the Third Reich.[7] Historians should not profit from others' misery.

First, *History* is a quasi-autobiographical work. It should be placed between Frankfurt Critical Theory and *Bildung*'s-materialism on the one hand, and Adorno's aesthetic philosophy of dissonance on the other. Kracauer tilled the field between the radical culturism of the early Frankfurt school influenced by Lukács, Weber, and Marx, and the aesthetic materialism of Georg Simmel. But any placing of his influences does not account for his essayistic practice of writing. The untheorized world of the essay is a place of research and praxis. Kracauer wrote in defiance of sociological or literary approaches to the contemporary.[8] For a generation raised on the photographic image and the film as the positive utopias of world unredeemed by any totality of perspective that would save the world from itself, it is no wonder that Kracauer becomes an iconic figure for having provided a series of short snapshot texts that in essence remind us of a world beyond the negative dialectics of Adorno or the indistinct "aura" of experience that hovers over Benjamin.

Second, Kracauer's work has not been understood as an intervention into the world, as theory as the practice of a Kantian moral geography where the essay speaks for the suffering of peoples who are left with a destroyed public sphere. His writing on film was an essayistic engagement with the publics' collective and cultural memory revealed in the individual's redemptive engagement with bourgeois culture in its totality. To put it metaphorically, film was the streets through which the spectators as groups wandered in search of the origins of their own anticipatory anxiety.

Individuality was formed in this anticipation of redemption around the corner. His essays are both interventions and guides into the foundation of that culture's complacency in regard to its own history and self-image—that is his writing is a genre breaking category of thought in action that began in the account of everyday culture building in the *Bildung's* axis of Berlin-Frankfurt genre of the *Feuilleton*. His meditative prose essays are not fragments, but *segments of a longer narrative*, which he sees reflected in the film-genre itself as the new form of authorship of *Bildung* in the age of psychology and mass media. Through hundreds of essayistic works he photographed memory and event (*Eindenken* und *Ereignisse*), Kracauer's materialist reading of reification and absorption into culture, revealed cultural transformations that the mind could not see.

Kracauer viewed the rise of the film as the opening to a new experiential category of being, therefore becoming a category of cultural creation. The film revealed the world stumbling on its path to cosmopolitan peace in an age of an-always-coming apocalypse. This cultural assimilation of totality was documented in *From Caligari to Hitler. A Psychological Study of the German Film* written in the United States but already underway in his Paris exile. His American work on film, *Theory of Film*, is on the other hand devoted to filmicly conceived cosmopolitan themes. *Theory of Film* documented the postwar American public sphere through the film form's capacity to *create* the spectator's viewing habits, which revealed the dominant social interests of the age. Here he located collective memory in the present, in the image of "now." The masses anticipate a future that would pacifiy reality.

And Kracauer wrote of the violence of the film's penetration into the group's consciousness of itself. The film becomes a genre-breaking essayistic form because of its capacity to redeem, save and participate in the illumination of physical reality through various heretical attitudes toward parvenu reality and the industrialization of the senses.[9] The film is a cultural form that realizes the hidden plan of our constructed nature, but the spectators—the new collective agents in modern life—were strangers everywhere and nowhere. They were nowhere at home in the postwar world. Empathic understanding of their collective immiseration and separation from each other and from culture was a requirement of a radical *Bildung* that in his Weimar essays illuminated the horrors that might emerge from the carnival of doubt and despair that underpinned the inward direction of their incomplete "salvation" through the public sphere *Bildung*. Already in 1930 Kracauer wrote to Adorno that with 3–4 million unemployed he saw no way out of a coming apocalypse. It wasn't only capitalism he wrote to Kracauer, but something "bestial" whose roots are deeper than capitalism. No salvation can be cranked up from the German people's unused power as was done in revolutionary Russia. The healing power of overthrowing the government was not be reckoned with.[10]

His *Feuilleton* essays, written before his ostracism in 1933, were careful autopsies of the economic and political failure of the republic to overcome corruption, charlatanism, and the servile and parvenu mentality stretching across the entire culture of officialdom. He anticipated the barbarism of the Nazis as well as the anti-Semitism and union-busting positions of the coming dictatorship. Politics were a carnival of despair. In 1933 he wrote to Benno Reifenberg from Paris (to whom he had dedicated *Die Angestellten*, his political ethnography of the social service workers and their mentality) that the unions might be the last bulwark against the coming Hitler dictatorship:

> My starting point is that the worst terror is expected here from a Hitler dictatorship, which will doubtless infiltrate even the smallest cottage. Do you believe that we would be spared in the event of a dictatorship? As far as it affects me personally, I am in all seriousness afraid that Jewish journalists will no longer be permitted to write. A movement within which the most furious hatred has been nourished for ten years will have its victims. Anti-Semitism, for us. All of this will be ruthlessly implemented, and it will eradicate German cultural life. Others have done the like before, and a country can survive without cultural life. . . . Assume that the horrible threats are carried out. We cannot trust the middle class, which allowed this movement to grow great, to controls its excesses. It is not much more than 100 years since German subjects (*Untertane*) were sold by their princes as soldiers. Think also of the withdrawal of passports. That can happen to us all.[11]

The cosmopolitan was a figure with two faces. On the one hand Kracauer understood the cosmopolitan as those who stood under the unsheltering heavens and who were "pariahs" in the sense that they reflected a conscious political identity of the inside-outside. At the same time, this persona was the carrier of the new urban phenomena that linked the bourgeoisie and antinationalist Jew with the masses of people collected under the sign of the strangers. The masses became performers in the group mentality that was being "collected" under the sign of the state. The social state and the class-strata, and the power-driven authorities were all part of the panorama of his Weimar essays. The conscious exile would become an emblem of modernity for the centralizing culture-states. However, unless the *Untertan* mutated into the historical stranger, the *Untertan* would remain the new exemplum, the proto-figure of modernity. Already a pariah in the nineteenth century according to Hannah Arendt, the stranger arrives on the doorstep of the modern city to be allowed in, but not to be allowed to stay too long. This idea was given shape in Simmel's classic essay of that name "The Stranger."[12] Kracauer discovers that the stranger is similar to the historian who lives between the micro and macro worlds of exterritoriality: "It is only in this state of self-effacement, or homelessness, that the historian can commune with the material of his concern [. . .] A stranger to the world evoked by its sources, he is faced with the task—the exile's task—of penetrating its outward appearances, so that he may learn to understand that world from within."[13]

The stranger who is evicted and then transplanted carries a modern existential displacement of a political condition. The stranger must now take on many forced identities; the stranger had before labored under the burden of paying a fixed tax to the powers, while the Christian citizens' tax burden "varied according to their wealth at any given time."[14] But now the stranger "intrudes as a supernumerary [. . .] into a group in which all the economic positions are already occupied."[15] The stranger is no longer a hybrid-cross cultural carrier of tolerance, generosity, and ethical friendliness. The stranger, for both Simmel and Kracauer, is freed from the pain of subjectivity by seeing the world in a disinterested fashion, dispassionately as an outsider. However, in Kracauer's original use of the film as a sign of cosmopolitan homelessness it is the nearness and farness of the film-created reality in which the emerging collective viewer resides, not burdened with any latent communal reciprocity or anticipatory longings that would have utopian bearings. The spectator of the film is the ultimate stranger in modernity who bears up under the cultural burden of time—the history of "now" time in which the illusion of peace sinks into the viewer's consciousness only to be revealed by the terror-filled image of photographic beauty.[16]

The new bearers of culture must overcome their shadowy existence in urban life in which communal existence had deteriorated, and where the spellbinding magic of the film foreshadowed the rise of the *new* collective mentality of a *Kulturstaat*. The new cosmopolitanism was a built-environment that contained a potential humanism grounded in Kracauer's acute sense of the communal, the local, the neighborhood, the street, the alleys, because modernity does not only produce fragmentation, but produces the segmented world of the "stranger as a member of a group" (Simmel):

> As such the stranger is near and far *at the same time* [*sic* Simmel's italics], as in any relationship based on merely universal human similarities. Between these two factors of nearness and distance however, a peculiar tension arises, since the consciousness of having only the absolutely general in common has exactly the effect of putting a special emphasis on that which is not common.[17]

This relationship of each to the group is erotic: the stranger is still an organic member of the group while "being inorganically appended to it."[18] The saving, redeeming, capitalist renewal in Germany would remove the appendage of culture, which would purify the organic body. The individual tarries in film.

Capitalism desecrates the communal through the imposition of an alienated otherness onto the minorities. The prophetic route that Kracauer followed from Simmel's world of the fragment into the anchoring of this world to the consciousness of the exile is given additional particularity by Kracauer's view of the Third Reich's barbaric qualities that could not be understood through economics or politics. Kracauer needed an approach that included cultural creation because *Bildung* was beyond the pale of the minorities and the racially impure, who must be violently annihilated in the name of national security. Kracauer's critique of culture not only draws its energy from Simmel, but from the spirit of the hidden stream of natural anarchism that emerged from the negative ontology of the incommensurable. This hidden stream is tapped in his last work, but flows throughout his earlier essays.

Kracauer searched for an antiauthoritarian position from which to open culture to the inner streams of yet-unorganized movements in a world of highly organized institutions. Kracauer's study of propaganda emerges from this antagonism toward the state's take-over of the public sphere. *Masse und Propaganda*, which was written in 1936, at a time when the Nazi judicially sanctioned terror had established its regime of anti-Semitic campaigns, marked the beginning of still another approach to film culture. The "spiritual shelterlessness of the masses" was wrapped in the collapse of the bourgeois hierarchy of value. The Bourgeoisie, lost in its self complacent-security, became problematic. This class was no longer continuous with the institutions that it had regarded as having withstood the contradictions of its alliance with capital. The capitalist interests come forward nakedly. Thus the purpose of Fascist propaganda is to reintegrate the masses into illusion. Terror and propaganda cannot be eliminated from the worldview of Fascism since the reign of terror is legitimated both in reality and illusion.[19]

While the sheltering institutions and the public sphere itself were falling apart, Siegfried Kracauer, from 1933 until 1941, carried on an intense, detailed almost weekly correspondence with Hedwig Kracauer, his aunt (his mother's sister). Behind the correspondence lies the sense of the doom of the exile. The correspondence details the daily lives, ideas and sense of the future in what was rapidly becoming a world of *Zeitgenossen* without *Genossen*, contemporaries without comrades. He had fled Germany at the first sign of the Nazi capture of power, and his letters hint at the fear, the despair and anger over the inability of the *Frankfurt Zeitung*, his employers, to help him when the Nazi pogrom was reaching west into France through the military offensives. He and his wife Lily became wards of the friends, libraries, publishers and emigration and visa authorities. Misbegotten plans, and the always impending destitution, desperate appeals to Max Horkheimer, Leo Löwenthal, Meyer Schapiro, and many others, made up their daily life. False hopes about securing posts in New York or Washington come monthly. He and Lily Kracauer leave Lisbon on April 15, 1941 and arrive in New York on April 25 with a few dollars, little English, and trunks of papers, letters, and manuscripts.

The correspondence frames his attempts to survive as a writer within the fugitive culture of Paris and remain free of the dread and fear of being hunted down as an intellectual, Jew, left-wing social critic and writer with clear critical, if not Marxist, tendencies. Hedwig Kracauer's futile attempts to organize a flight from Frankfurt for another country is one of the themes of their correspondence as is the discussion of films, birthdays, and relatives. His attempts to meet them or rescue them are abandoned many times. Planning is the script. They are the players. The failure to rescue them and to protect them from their illusions of a normal everyday life in Germany and their blindness to their fate becomes, in my reading of his *Caligari, Theory of Film, and History. The Last Things Before the Last* the inner source of his method of combating the social death that accompanied his optics of contemporaniety.

During this Fascist-created exile, he saved his mind and soul by grounding himself in the world of a pre-Fascist exterritoriallity in Weimar, that inspired him to complete his Offenbach book, *Orpheus in Paris. Jacques Offenbach and his Times*, as well as a major study on propaganda, always while researching films. He planned other projects, including a film script. The study of Fascist propaganda (1936) was the genesis of *Caligari*, but more importantly it established the fugitive's aesthetics that characterized an exterritorial "elsewhere" in film and in letters that would continue to reflect his historical understanding of the relationship of political exile to the ongoing world-historical dispersion of peoples that seemed to be the wave of the future. Exiles, refugees, statelessness, anarchists, deportees, communists, working people and "crowds of people" "hungering for life" and who are deprived of the historical stream of the "breathing world [. . .] that stream of things and events which, were it flowing through him would render his existence more exciting and significant."[20] It is in *Theory of Film* that his "colportage" sense of history became truth-to-life, in the way *Caligari* was truth to the past.

Kracauer's letters to his aunt and mother observe exile as symptoms of history. He writes as a witnessing therapist about the competing, socially stratified chauvinistic movements incubating in a culture of ressentiment among members of the intelligentsia and the emerging mass audiences. She writes of the daily lives of the Frankfurt intelligentsia. This becomes the backdrop to the Offenbach study. *Offenbach* brings to life a figure who lives in an imaginary territory with no centre. Nineteenth-century Paris is a resettlement camp-city, a boulevard for the European peoples who are creating new identities without a common or homogeneous tradition. The wars of the Habsburgs released the individual for a new medium of entertainment Offenbach creates a new idiom that breaches cultural borders by particularizing central European economic despair in his librettos and music. Kracauer particularizes and narrates— relives—this despair in the same way he reads the Weimar films: as biographies of the age and photographed history. The dispersal of the European peoples is forecast in the genre of musical-theatre. It is no *Dreigroschen Oper*, but Kracauer nevertheless heard the sounds of migration and homelessness, imprisonment and the catastrophic state violence in the genre of the "operetta, [which] would never have been born had the society of the time not itself been operetta-like; had it not been living in a dream-world, obstinately refusing to wake up and face reality."[21]

Offenbach and *Caligari* are "exilic texts," Orphic texts, that is texts that focus on insecurity and demonic depths inside the subject matter. They are allegories of power and despair for the new entertainment elites that use mass culture as their new homeland. *Caligari* insinuates biological racism, discredits intellectuals, demonizes women and weak fathers, who are all united by their puritanical reinforcement of sensual images of defeat under the star of the relativity of all social movements. *Offenbach* approached the same subject in another way: through a boulevard artist who exploits his own weakness in order to lay "bare the dark foundations out of which operettas had grown [. . .]." [22]

The Weimar film harbors the deep allegorical optic view of broken ethical principles now reduced to moral homilies. Images of defeat tapped into the darkest fears of society on the road to depoliticization. The naturalness of culture, which might be an antidote to class warfare, was crushed again and again by the contradictions in capitalist formations. Weimar film seen through the lens of Kracauer's exile is a cultural form that expresses itself as an avant garde ruin. The film becomes archaic as soon as it is assimilated into the culture. To use the metaphor of architecture and apply it to this exilic text, *Caligari*-films express what Hitler and Speer hoped to bring to the monumental buildings they fantasized: in 1000 years their architecture would look like ruins. The employment crisis had been portrayed in Kracauer's pathbreaking book *Die Angestellten* (1930) as an asylum for the unsheltered. At the same time he saw in the origins of German film a ripening nostalgia for a feudal world that always lurked around the corner. *Caligari* the film staged an oedipal drama of the age. The age masked the blood and ethnicity of its politics in the relentless lust to form a national identity that appeared to wane and ebb and return at every tick of the historical clock.

The Archimedean U Turn to Reparation, *Bildung*, and Inscape

In *History. The Last Things Before the Last* Kracauer turns back to the Kafka he wrote about in the 1920s. In one of his last essays before the flight to Paris he quoted Kafka's "The City Coat of Arms": "All the legends and songs that came to birth in that city are filled with longing for a prophesied day when the city would be destroyed by five successive blows from a gigantic fist. It is for that reason too that the city has a closed fist on its coast of arms." Kracauer concluded this essay by writing: "With the unconfirmed longing for the location (*Ort*) of freedom we stay here." But he did not "stay here." In his last work he investigated the dead to recover them from the "frozen time" when escape from certain death cast a shadow over the already shadowy ports of call in his and Lily Kracauer's journey. Any "last thing" for the memory of survivor-history lies in the work of those who come afterward where the longing (*Sehnsucht*) for an aesthetic and historiographical truth would be embedded in the exile consciousness that overwhelmed a large part of the world. He writes in *Theory of Film* about "phenomena overwhelming consciousness" that "Elemental catastrophes, the atrocities of war, acts of violence and terror, sexual debauchery, and death are events which tend to overwhelm consciousness. In any case, they call forth excitements and agonies bound to thwart detached observation. No one witnessing such an event, let alone playing an active part in it, should therefore be expected accurately to account for what he has seen."[23] The cinema, he writes, "aims at transforming the agitated witness into a conscious observer."[24] The intellectual and critical diaspora spoke for themselves in the anteroom where elapsed time would, if taken as a large narrative of general history, have exhausted the energies in yearning and mourning for those dead and for unrepaired images of those who would come after.

Behind the *Theory of Film* lies a work of mourning that was yet to be written. The work in sight throughout the last years of his life turns out to be one of the most astonishing literary works of the time. Unclassifiable, novelistic, meditative, and yet with a sense of timing, which was always part of Kracauer's genius, *History. The Last Things Before the Last*, is a colportage study of how history forms itself in the time-space of one's memory. His notebooks for this study clearly evolve from his contact with intellectuals and academics, art historians, and cultural figures whom he had advised or assessed in his work with the Bollingen Foundation, Museum of Modern Art. The argument has been made that Kracauer's language is "metaphorical."[25] This is not the case at all. His language is iconic, that is he spatializes authorship and teaches the addressee in a noncoercive manner, illuminating the nearness and farness of his architectural framework in writing about immediate life. Put more directly, he does not use metaphor, montage, or a series of quotations in the manner of Benjamin; he stands more directly in the path of memory by photographing memory through the lens of the author in the work. Borrowing from Kafka he uses fantasy and judgement in the manner that recalls his study of Kant.

In Paris Kracauer had already gone through a state of anticipatory mourning about loss of friendship, relatives, language, profession, image, and shelter—the state of being destitute and *Obdachlos* (shelterless). Exterritoriality is not a metaphor—it is an image that displaces all metaphor by photographing the state of fleeing from German killers who murdered the intellectual life of a generation. It becomes the root icon for his writing that depicted a social system dependent on reification and exchange of the person for abstract powers outside the realm of nature. In a letter to Max Horkheimer he wrote (1941):

And the thought of suicide was also not far at a certain time. We were, incidentally, often together with Benjamin in Marseilles, and we hiked over the mountains to Port-Bou only a few days after him, where we were turned away, like him. It was the darkest time of eight dark years. If I can get halfway settled down, I will write down our experiences (*Erlebnisse*) during the last two years, under the title of "Voyage to America." I shall either succeed in putting Kafka in the shade, or I was incapable of properly presenting the events.[26]

He encounters memory and photography suddenly, stumbling into a photographically illuminated, black–white Kafka-like attic. He discovers there an album in the bottom drawer of a trunk. This event is not hard to imagine, for he actually did—in 1950—open some long closed trunks. He wrote to T. W. Adorno that he had found some lost papers and that he might now write his memoirs. This yearning to account for his life was never actually carried out. Instead looking in the bottom drawer of the trunk of his considerable files of frozen speech, which he had borne with him through after 1933, he found that he would have to somehow integrate his past with his present around the idea of the photograph. As Paul Kristeller, who wrote the introduction to the posthumously published study, notes, this work was a kind of moral geography about his own life. It was a past, which at seventy-seven, he was only able to formulate through a kind of historiographical series of iconic images that allowed him to remember his earliest creative years before the flight from certain death. Even though *History* was conceived as a self-witnessing, and was by and large written meticulously on copious note cards and carefully articulated through his visually sensitive architectonic mind, it was always interrupted by his work with the Rockefeller and Bollingen Foundations and the desire to see his other exile writings translated into German. Unlike Adorno, whose writing contains a sense of leisure only by combating in an agitated, dictated prose any lockstep idea of sequence, Kracauer worked arduously on his projects, always distracted from completing them. Reparation was a blueprint. In 1949 in a letter to an old friend, Bernhard Guttman, Kracauer wrote:

> For a long time, I remained as if lame whenever it came to renewing beloved relationships, which inevitably opened wounds of the past. My old mother and aunt were deported early in the war from Frankfurt to Theresienstadt. In 1945, we learned that they were dragged from there to Poland. That is all we know, and that suffices for our lifetime.[27]

This terrible thought is reflected in *History* that the "historian is not just the son of his time in the sense that his outlook could be defined in terms of contemporary influences [. . .]. Nor is his conception of the past necessarily an expression of present interest, present thought; or rather, if it is, his aggressiveness may cause the past to withdraw from him. The historian's mind is in a measure capable of moving about at liberty."[28] As if commenting on his own life after Terezin and Auschwitz, he continues: "And to the extent that he makes use of this freedom he may indeed come face to face with things past."[29] What is this freedom? It is the freedom of creating reparation for the damages that history creates. The issue is not whether there could be culture or poetics after Auschwitz, but could there be history as such? The world that was divided into places where the historian may not go, and where the Jews could not live, must be places where the historian can go even if the pariahs may not enter there. This history must work through the nature of the *many* genres of exterritoriality, that is exilic texts. This thought may be close to Benjamin's and Adorno's idea that in all culture there is a little barbarism, yet in *History*, Kracauer separates himself from both by establishing the reality of himself as a Kafka man of good will who takes the reader on a journey that is not in any sense of the word apocalyptic or eschatological. Not content with historicizing through general history, which in his Weimar essays he would have labeled with other terms or influences or structures of social movements, he uses without the slightest pause or bridge a different discourse: "Orpheus descended into Tartarus to fetch back the beloved who had died from the bite of the serpent." The "plaintive music" spoke "that Orpheus might not look behind him until she was safely back under the light of the sun [. . .]. Like Orpheus, the historian must descend into the nether world to bring the dead back to life [. . .] and what happens to the Pied Piper himself on his way down and up? *Consider that his journey is not simply a return trip [my italics—JZ]*."[30] This is an unguarded reference to the death of his mother and aunt and the dead souls who lie in the interstices of history.

The essay on history requires an almost line by line analysis of this meditation on memory and reparation, because his almost mathematical reasoning about exiled history unfolds in an architectonic, iconic form—that searches for an "optical" projection of what a life history would be if seen from the point of view of the public world. In the idea of the "the last thing before the last," Kracauer is making peace with himself, and his memory of the dead souls were being bartered away in his American exile where he had to avoid accommodation to the world of America. Like Kafka's Karl Rossman, Kracauer vanished into America as a nobody. Both Karl and Kracauer and became visible only slowly and consciously to themselves and others.[31] Until his flight in 1933, Kracauer had produced almost 2000 essays and commentaries and now at seventy-five he needed a panoramic-helicopter view that would enable him to see that these essays cohered as a personal myth. The historian must follow Kafka, and like Ranke, "blot out his self, not just for the purpose of dispassionately rendering the course of past events but with a view to becoming a *participant observer* engrossed in the uniquely significant spectacle that evolves on the stage of the world."[32]

What is this ageless state of near and far optics? It is "the near-vacuum of extra-territoriality, the very no man's land which Marcel entered when he first caught sight of his grandmother?"[33] Here the shadow of the Proustian–Kafkaesque object falls over both the memory and image of his mother and aunt. Yet they remain nameless. What is this emotional detachment, which Kracauer compares favorably to the photograph, especially when he enlists Proust's Marcel who sees his grandmother as she really is?—"His inner picture yields to the photograph at the very moment when the loving person he is shrinks into an impersonal stranger; as a stranger he may indeed perceive anything because nothing he sees is pregnant with memories that would narrow his field of vision."[34]

In the final sections of this meditative essay on reparation, Kracauer came to his objective: the redefinition and rehabilitation of cultural history as the only reliable framework that would allow us to avoid the owl of Minerva's flight into dialectical overdetermination—Adorno's disease. The cast of characters are not Germans nor the German language—they are fugitives from their society. They carry the burden of culture—not traditional culture but the culture of phenomena as actualities in their own right. This, says Kracauer, requires the energy of a Kafka who lives within the "non-homogeneous structure of the intellectual universe."[35] He places Kafka directly against both philosophy as the queen of truth-value statements, and against Adorno's anti-philosophy that loses its materiality in the way dialectics accommodates itself to the world and coerces the dialogical into exile. Although this may seem like a philistine's critique of the great Adorno, it is grounded in Kracauer's return to his own lifeworld of the in-between. He cites C. Wright Mills's sociological handwork; he looks into the world of Huizinga, where the resurrection of the object takes place in the "sphere of the dream, a seeing of tangible figures, a hearing of half-understood words." He writes:

> Taken with Huizinga's rehabilitation of the antiquarian interest in the same context, this is one of the best definitions I have come across so far of the historians breakthrough into the realm of ideas. It shows Huizinga's awareness that his ideas transcend subjectivity even though they grow out of it, and thus disposes, at one stroke, of the quarrel about the *share of subjectivity* in the historian's knowledge. Indeed, a stopping mid-way may be the ultimate wisdom in the anteroom. Hence throughout this book my concern with shades and approximations.[36]

This "impressionistic" turn to hues, mood, shades, approximations is as much an architectonic as a psychological enactment. It is motivated by the photograph's capacity to stimulate memory, but it is the mindful "mood" of the exile who lives on the edge of loss and mourning, and yet is outside the systems of tragic knowledge that cast their shadows on the nameless ones who wander the world as stateless beings. In *History* there is an anarchist's restlessness about those who live outside of the state. Here are the words Kracauer uses: nonhomogeneous; sphere of a dream; rehabilitation; share of subjectivity; stopping mid-way; approximations; nameless; hiddenness; interstices, and finally, quoting Hans Blumenberg, he finds another ally in finding hidden possibilities as the basis of a future anarchistic ethics of the face-to-face. History is not a charnel house it is utopic. Memory is utopic like the modern film because memory is optical and narrational. It is conditioned by time-space, it is chronotopic, and never leaves us, it is our body based insight that history cannot fully capture; memory must show the highest form of truth in the porosity of the natural history of forgetting that the surfaces recall.[37]

There can be no culture without memory even if collective memory counterfeits general history, Kracauer praises the Auerbachian realism of Woolf, Joyce, Proust, or Valery who do not mistake "ancestry" for allowing us access to a method of relating experience. General history "then, is a hybrid something between legend and the Ploetz, that imperishable annalistic manual which we as schoolboys use to memorize the dates of battles and kings."[38] The turn to photography is a return to the Kantian aesthetic of engagement with a rights-based aesthetic of the person. This *Bildungs*-aesthetic lies between "the hazy expanses in which we form opinions"[39] and the "lofty aspirations," which are the objectives of a progressive culture—the peculiar area of "an intermediary area which has not yet been fully recognized as such." In this intermediary area lies Kracauer's sense of the terms of a hopeful future. Memory of "pockets and voids"[40] that lie "both inside and outside flowing time."[41] He asks: "Is their inherent claim to temporal exterritoriality justified?"[42] or does their share in historical relativity invalidate any such claim?" The "binding power" of today's alienating realities "would still be limited by the mind's freedom to initiate new situations, new systems of relationships."[43] The idea that the effects of an age determine whether "people actually 'belong' to their period. This must not be so. Vico is an outstanding instance of chronological exterritoriality [. . .]. Like great artists or thinkers, great historians are biological freaks: they father the time that has fathered them. Perhaps the same holds true of mass movements, revolutions."[44]

If exile is the state of compromising solitude with isolation, why do readers find the book so equivocal? Kracauer is very clear: it means imposing one's self-alienation onto one's inscape, not as a mental landscape but as in Kafka's "strange mixture of hopelessness and a constructive will which did not in his case cancel each other out but by their wrestling with each other were driven to rise to infinitely complicated expressions" (Kracauer quoting Max Brod).[45] A spirit of noplace has been displaced, and in its place we have the discovery, in the phenomenological sense, of the temporal character of existence, which can only be grounded in substance when the symptoms— or mythos—of the age are understood methodologically through bringing about the "visibility" of what is hidden. The age shows itself in the way groups and masses, crowds and powers do not only reveal alienated states of powerlessness, but the ways in which solidarity and friendship, and the "active passivity" of the historiographer–story teller.

The shadow of the object of reparation continued to pursue this writer through-out his exile. He realized very late in life that he existed in the interstices of history, virtually forgotten. His fate was to witness the materiality of this earthly existence, Kafka-like, and also to reflect on those who had caused and then allowed and then watched and did nothing while something terrible was happening.

Postscript: The Last Exile before the Last—The Photographable Present

There will always be exile if we believe Kracauer's final testament. What can we learn from his fate? Is he "something for us" in our own existence in the interstices as we— "focus on the 'genuine' hidden in the interstices between dogmatised beliefs of the world, thus establishing a tradition of lost causes; giving names to the hitherto unnamed?"[46] In the "Lieu of an Epilogue" attached to the final pages of *History*, Kracauer quotes Kafka's parable "Concerning Genuineness," which is a parable also behind Benjamin's "Angelus Novus," the angel of history. Kracauer's angel however looks to the "Utopia of the in-between—a terra incognita in the hollows between the lands we know."[47] It is not an eschatological view of the future, it is the fulfillment of reparation, the escape from death. "What photographs," Kracauer wrote in 1927, "by their sheer accumulation attempt to banish is the recollection of death, which is part and parcel of every memory image. In the illustrated magazines the world has become a photographable present, and the photographed present has been entirely eternal-ized. Seemingly ripped from the clutch of death, in reality it has succumbed to it."[48]

Today Kracauer is not completely forgotten although no commentator has really studied him as the founder of a genre of essay of reparation that engages the public and private spheres as a quasi-anarchist spirit who used film and photography for the pursuit of a topography of collective consciousness. He commented many times that he was not a film theorist; he used film to prepare the world to see, what Merleau-Ponty described in "Cezanne's Doubt":

> But let us make no mistake about this freedom. Let us not imagine an abstract force which could superimpose its effects on life's "givens" or cause breaches in life's development. Although it is certain that a person's life does not *explain* his work, it is equally certain that the two are connected. The truth is *that that work to be done called for that life.*[49]

What can we gain from his legacy in a world grown more overwhelmed by a globally determined capitalism? Perhaps to realize that exile as a subject must be connected to the state of the fate of the republic of contemporary refugees, which is the modern equivalent of the exterritorial state. We can if we wish witness, through the window of Kracauer's "principles of history": people living in camps, zones, enclaves, ghettos that are "governed" by refugee commissions and UN agencies. Hidden in the interstices of the borders of this world there have been twenty million displaced persons over the last twenty years—a population larger than Australia and almost as large as Canada. This constitutes thirty-seven people per square mile and if imagined as a state-territory would be the size of France, Germany, England, and Italy. This exilic-culture is contemporaneous with borders and wars at the borders. This "state" is not charted or mapped. The sparsely populated areas of the United States, Argentina, Canada, Brazil, and Australia historic immigration lands are mapped. There is no capital, no representation, no real territory, no democratic structure. If this republic of exile were a state it would be in the top 10 percent of the states, above Turkey and below Italy. Here are the multilingual, multiethnic wards of international agencies. We are living with a situation that may be too much like the last two years of World War II when, as Michael Marrus has pointed out, there may have been as many as thirty million refugees in Europe two years before the end of the war.[50] At the end of the war seven million souls lived exterritorially in the Allied zones and an equal number in the Soviet. No one escaped being identified as a "refugee." Exile studies today as in the postwar period must be marked a politics and ethics of recognition of statelessness—of the vast numbers of refugees who live with the same illusions that they are being seen, and heard and will be saved by the promises of economic and legal freedoms. This is Kracauer's legacy, prefigured in Kracauer's small essay on Kafka in 1931 about the "Coat of Arms," which makes historical sense as prophecy and reparation the memory of exile and how exile and the fear of exile is inside our culture. Kracauer's own "coat of arms" is a plaque (Erinnerungstafel) installed in 1994 at Sternstraße 29, Frankfurt where he and his family lived from 1919–1930: "ER WURDE 1933 INS EXIL GETRIEBEN" (he was forced into exile 1933) with the epigraph continuing in capital letters from Kracauer's "Gestalt und Zerfall": "DENN DAS GESTALTETE KANN NICHT GELEBT WERDEN, WENN DAS ZERFALLENE NICHT EINGESAMMELT UND MITGENOMMEN WIRD."[51]

Notes

1. Siegfried Kracauer, *History. The Last Thing Before the Last* (New York: Oxford University Press, 1969), v (hereafter referred to as *History*).

2. See my article "Siegfried Kracauer's Cosmopolitan Homelessness—The Lost Cause of an Idea in the Film Age," in Suszanne Kirkbright (ed.), *Cosmopolitans in the Modern World. Studies on a Theme in German and Austrian Literary Culture* (München: Judicium, 2000).
3. Quoted in Matthias Wegner, *Exil und Literatur. Deutsche Schriftsteller im Ausland, 1933–1945* (Frankfurt, Bonn: Athenäum, 1968), 161.
4. Carolyn Forche, ed., *Against Forgetting* (New York: W.W. Norton & Co, 1993): 51.
5. See my essay "Decamouflaging Memory, or, How We Are Undergoing 'Tria by Space' While Utopian Communities Are Restoring the Powers of Recall," *West Coast Line 34/35* (Spring 2001): 119–157.
6. The phrase is from Siegfried Kracauer, *Theory of Film. The Redemption of Physical Reality* (Oxford: Oxford University Press, 1960), 287 (hereafter cited as *Theory of Film*). Miriam Hansen's foreword to Alexander Kluge and Oskar Negt, *Public Sphere and Experience. Toward an Analysis of the Bourgeois and Proletarian Public Sphere* (Minneapolis, London: University of Minnesota Press, 1993; German edition, 1973), is an exception since she sees Kracauer's work as political in the sense of Kluge's and Negt's.
7. Henry Pachter, "On Being an Exile," *The Legacy of the German Refugee Intellectuals. "Salmagundi"* 10–11 (Fall–Winter 1970): 14.
8. See Adorno, "The Essay as Form," *New German Critique* 32 (1984).
9. Kracauer, *Theory of Film*.
10. (Letter to Adorno, August 28, 1930, Deutsches Literatur Archiv [hereafter cited as *DLA*]).
11. Kracauer to Reifenberg, February 18, 1933, *DLA*.
12. Whereas Simmel's "stranger" was caught in the in-between world of inside–outside, Kracauer went beyond Simmel. Also see another essay written by Kracauer's fellow exile, Alfred Schuetz, "The Stranger: An Essay in Social Psychology," *The American Journal of Sociology* 49 (1944): 499–507. Schuetz writes to Aron Gurwitsch, who was himself interned with the Kracauers in the Gurs camp, that Kracauer was threatening suicide (letter of November 2, 1940, p. 19). An excellent analysis of Arendt's pariah-parvenu thesis can be found in Anson Rabinbach, "Hannah Arendt writes 'The Jew as Pariah: A Hidden Tradition,' in which she describes the forgotten tradition of Jewish 'conscious pariahs,' " in *Yale Companion to Jewish Writing and Thought, 1096–1996*. ed. Sander L. Gilman and Jack Zipes (New Haven, London: Yale University Press, 1997), 606–613.
13. Kracauer, *History*, 84.
14. Georg Simmel, "The Stranger," in *On Individuality and Social Forms*, ed. Donald N. Levine (Chicago, London: University of Chicago Press, 1971), 149 (hereafter cited as *The Stranger*).
15. Simmel, *The Stranger*, 144.
16. This mistaken view of Kracauer as a nomad or, worse, a rootless transplant into America, persists in Enzo Traverso. First given sentimental standing by Walter Benjamin's characterization of him as an *"Außenseiter"* and *"Lumpensammler"* (an outsider and "ragpicker in the dawn of the revolution"), *Ein Außenseiter macht sich bemerkbar*, in Walter Benjamin, *Gesammelte Schriften*, vol. 3 (Frankfurt: Suhrkamp, 1991), 219–225 he was almost codified into oblivion by T. W. Adorno's "The Curious Realist," which Kracauer bitterly rejected for good reason. This Adornoesque view persists in those views that fail to see how Kracauer used and negotiated his exile condition into a creative-critical utopian view of the film as a form of judgement and exit from our own exile in this world. He is characterized as a "nihilistic utopian" (Heide Schlüpmann), permanently embedded in exile (Karsten Witte), ontologically exiled and permanently unhappy (Gertrude Koch) and overwhelmed by "the fear of inescapability" (Mülder-Bach). From the beginning of the Kracauer revival, initiated in the generation of the *Nachgeboren* by Martin Jay's astute insights into Kracauer's self-analysis of his own "exterritoriality," this position has the effect of denaturing his essayistic critical approach to "totality" and redemptive anarchism that comes to fruition in the *anti-messianic post-Terezin/Ausschwitz* plea for an open non redemptive conception of history *History*. See: T. W. Adorno, "The Curious Realist: On

Siegfried Kracauer," *New German Critique* 54 (Fall 1991), originally a radio address and then an essay in T. W. Adorno, *Noten zur Literatur* III (Franfurt: Suhrkamp, 1966); Enzo Traverso, *Siegfried Kracauer, Itinéraire d'un intellectuel nomade* (Paris: éditions la découverte, 1994); Heide Schlüpman: "The Subject of Survival: On Kracauer's *Theory of Film, New German Critique* 54 (Fall 1991); Karsten Witte, "*Siegfried Kracauer im Exil,*" *Exilforschung* 4 (1987): 135–149; Gertrude Koch, *Siegfried Kracauer, An Introduction*, trans. Jeremy Gaines (Princeton: Princeton University Press, 2000) [translated from *Kracauer zur Einführung*, Hamburg: Junius Verlag, 1995); Inka Mülder-Bach, "History as Autobiography: The Last Things Before the Last," *New German Critique* 54 (Fall 1991). Of Martin Jay's several astute essays on Kracauer, the one referred to here is "The Extraterritorial Life of Siegfried Kracauer," *Salmagundi* 31–32 (Fall–Winter 1975–1976): 49–106. See also "Adorno and Kracauer: Notes on a Troubled Friendship," in *Permanent Exiles, Essays on the Intellectual Migration from Germany to America* (New York, Columbia, 1985).

17. Simmel, *The Stranger*, 148.
18. Ibid., 149.
19. *Exposé Masse und Propaganda. Eine Untersuchung über die fascistische Propaganda. DLA* unpublished mss. 1936. The existence of the full manuscript is not known, but what is known is that Kracauer was deeply offended by Adorno and Horkheimer's editorial emasculation of the manuscript when they demanded alterations that Kracauer refused to do.
20. Siegfried Kracauer, "Hunger for Life," in *Theory of Film*, 167–169. Kracauer makes a specific point of writing that the depiction of the terror of the concentration camp that is "too dreadful to be beheld in reality" in the horrorhouse of film allows us to "redeem horror" obliquely. His example is the depiction of the slaughterhouse of methodical killing of animals in Georges Franju's *Le Sang des Betes*, 305–306.
21. Siegfried Kracauer, *Orpheus in Paris. Offenbach and the Paris of his Time*, trans. Gwenda David and Eric Mosbacher (New York: Vienna House, 1972), 185.
22. Kracauer, *Orpheus in Paris*, 363.
23. Kracauer, *Theory of Film*, 57.
24. Ibid., 59.
25. See Gertrude Koch, "Exile, Memory, and Image in Conception of History," *New German Critique* 54 (Fall 1991). Koch is a perceptive reader of Kracauer but like many critics of Kracauer she errs in her interpretation of him as one who "blurs" or "vacillates" or flight into the "imaginary."
26. Kracauer to Horkheimer, *DLA*, June 11, 1941.
27. The key word here is "dragged"! *DLA*, Kracauer to Bernard Guttman, August 8, 1949.
28. Kracauer, *History*, 79.
29. Ibid., 79.
30. Ibid., 79.
31. Kracauer wrote about Kafka's *America* (*Der Verschollene*—he who vanishes) in the *Frankfurer Zeitung* (23.12.1927). He wrote three other substantial essay-reviews: "Der Prozeß" (The Trial), *Frankfurter Zeitung* (1.11.1925); "Das Schloß" (The Castle), in *Frankfurter Zeitung* (28.11.1926); and a long essay, "Franz Kafka, Zu seinen nachgelassenen Schriften," *Frankfurter Zeitung* (3 and 9.9.1931), which would have been one of the last contributions before his exile, which ends with a poignant commentary on Kafka's "The City Coat of Arms" that I have quoted. These essays can be found in *Franz Kafka, Kritik und Rezeption. 1924–1938*, ed. by Jürgen Born (Frankfurt: Fischer, 1983).
32. Kracauer, *History*, 81.
33. Ibid., 83.
34. Ibid., 83.
35. Ibid., 212.
36. Ibid., 213.
37. Ibid., 181.
38. Ibid., 190.

39. Ibid., 191.
40. Ibid., 199.
41. Ibid., 200.
42. Ibid.
43. Ibid., 66–67.
44. Ibid., 69.
45. Ibid., 216.
46. Ibid., 219.
47. Ibid., 217.
48. Kracauer, "Photography," in *Mass Ornament*, 59.
49. Maurice Merleau Ponty, "Cezanne's Doubt," in *The Merleau-Ponty Aesthetics Reader: Philosophy and Painting* (Evanston: Northwestern University Press, 1993). Italics Merleu-Ponty.
50. Michael Marrus, *The Unwanted—European Refugees in the Twentieth Century* (New York: Oxford University Press, 1985), 299. These startling figures have been culled from various U.N.H.C.R. (United Nations High Commission for Refugees) reports and were displayed in a photographic installation "Surveillance" at the Prague Castle in Prague, 2002 by Ingo Günther and Stefan Matthys entitled "Refugee Republic."
51. Translation: "For what has been formed cannot be lived unless what has fallen apart is gathered up and taken along."

CHAPTER ELEVEN

HORKHEIMER, ADORNO, AND THE SIGNIFICANCE OF ANTI-SEMITISM: THE EXILE YEARS

Jack Jacobs

When, in 1934, Max Horkheimer, Director of the Institute of Social Research, went into exile in the United States, he was a Marxist.[1] By the time that he returned to Frankfurt am Main, in 1949, Horkheimer had distanced himself considerably from Marxism of any kind.[2] This general tendency in Horkheimer's work was particularly evident in Horkheimer's writings on anti-Semitism.[3] It was, moreover, also clear in the work on that subject done during the exile years by many of Horkheimer's closest collaborators in the Institute of Social Research. The reassessment of the nature and significance of anti-Semitism made by the Critical Theorists while in exile may have contributed to the overarching alteration of Critical Theory itself. The changes in Horkheimer's approach to anti-Semitism, however, were linked not so much to his American experiences as an exile as to the growing influence of Adorno and to the Institute's recognition of events in Europe.

As late as 1938, Horkheimer could still write a bluntly Marxist article on the Jews and Europe. "Whoever is not willing to talk about capitalism" Horkheimer famously proclaimed in this article "should also keep quiet about fascism [. . .]. As agents of circulation the Jews have no future. They will not be able to live as human beings until human beings finally put an end to prehistory."[4] Theodor W. Adorno voiced no reservations about Horkheimer's argument.[5] Indeed, when Adorno described the article to Walter Benjamin he reported that he had "intensively collaborated" on it.[6] However, it was ultimately Adorno who led Horkheimer to revise his understanding of the persecution of the Jews in Europe, and to analyze anti-Semitism in a manner more consistent with Horkheimer's emerging approach to other issues.

In a prescient letter written by Adorno to Horkheimer on February 15, 1938— that is, on the day before Adorno left Europe for the United States—Adorno predicted that there would be no country that would admit Jews who remained in Germany, and that those Jews would, therefore, be extirpated [*ausgerottet*].[7] This, it must be underscored, is not only before the extermination of the Jews had actually begun, but also before most people—including most Jews in Germany—believed that such extirpation would actually take place.[8] The timing of Adorno's premonition is significant, for it demonstrates both that Adorno was thinking about anti-Semitism in a dramatic manner on the eve of his arrival in the United States and that Adorno thought that anti-Semitism was a vital issue to raise with Horkheimer. Adorno continued to concern himself with the fate of the Jews, and the significance of this fate, for years to come.

Neither Adorno's premonition of 1938, nor his increasing conviction of the importance of anti-Semitism, can be attributed to Jewish identity as such. Though Oskar Wiesengrund—Theodor Adorno's father—had been born a Jew, he had converted to Protestantism, and had allowed Theodor to be baptized.[9] Theodor's mother was a Catholic, and, in 1924, Theodor himself (who had been confirmed in a Protestant church) had considered converting to Catholicism "die mir als dem Sohn einer sehr katholischen Mutter nahe genug lag."[10]

Adorno is not known to have evinced any particular interest in Jewish affairs during the 1920s, was not attracted to the theoreticians of the Jewish cultural Renaissance, and was not directly involved with any explicitly Jewish organizations. When his friend, Sigfried Kracauer told Adorno about Rosenzweig's *Star of Redemption*, Adorno quipped "These are linguistic philosophemes I would not understand even if I understood them."[11] Several years later, when Lowenthal and Erich Fromm (both of whom had taught at the Lehrhaus, and both of whom had been intimately involved with Judaism and with Jewish organizations prior to their affiliation with the Institute of Social Research) became members of the Institute, Adorno referred to them as "Berufsjuden," a term dripping with disdain.[12]

There is no evidence that Adorno was ever subjected to anti-Semitism as a young man. Indeed, years later Adorno claimed that anti-Semitism had been "quite unusual in the commercial city of Frankfurt," the city in which he had been born and raised.[13] The success of the Nazi movement, of course, dramatically altered Frankfurt's atmosphere. In September, 1933, Adorno lost his official permission to teach because he was a "non-Aryan."[14] However, he initially responded to this change not by publicly condemning anti-Semitism, but rather by "hibernating."[15] When, later in 1933, the composer Alban Berg complained to his one-time student Adorno that not a single note of his was being played in Germany, the fact that he was not a Jew notwithstanding,[16] Adorno replied:

> In your place, I would in any case write . . . to the Music Chamber and make it clear that you are . . . of purely Aryan origins. As long as the Hubermans and Walters are celebrated in Vienna and you are forgotten, I cannot see why you should maintain solidarity with a Jewish population [*Judenheit*] to which you do not belong, which surely maintains none with you, and about which one should harbor as few illusions as about other matters. It may perhaps interest you in this connection that the "Jewish Cultural Association" that was initiated with such fanfare in Frankfurt has refused me participation because I am of Christian confession and only a half-Jew by race.[17]

In 1934 (when, to be sure, some might say there were extenuating circumstances) Adorno went so far as to write a review for a periodical that had become ever more sympathetic to Nazi views, and which eventually became an organ of the Reichsjugend.[18] "[. . .] I [attempted] to remain in Germany, at any price [. . .]"

Adorno wrote in 1934, "When it just wouldn't work any more and I was deprived of one possibility after another, even the most modest—even in giving music instruction I was restricted to non-Aryans—I decided to leave, despite everything (by the way, I could easily have survived materially in Germany, and I would have had no political objection either).[19] Even if Adorno's contribution to a pro-Nazi periodical can be rationalized and his self-proclaimed "lack of political objection" to remaining in Germany dismissed as hyperbole, they seem to corroborate the sense that Adorno did not think of himself as Jewish during the Weimar years or at the beginning of the Nazi era. Yet, as Benjamin was to point out to Adorno after World War II had begun, "The very fact that Proust was only half Jewish allowed him insight into the highly precarious structure of assimilation; an insight which was then externally confirmed by the Dreyfus Affair."[20] Much the same can be said about the source of Adorno's insights into anti-Semitism, later confirmed by the Holocaust.

Adorno had not been employed as a regular member of the Institute of Social Research during the Weimar era. Indeed, he did not have such a position until 1940, and it was not until this long-hoped for position finally materialized that Adorno and Horkheimer could devote themselves, as Adorno had hoped for quite some time, to the project of cowriting a major work.[21]

The precise topics around which this joint work would revolve were, for an extended period, still very much up in the air. Horkheimer was busy attempting to acquire funding for any number of different research projects, on any number of different topics—including a project on anti-Semitism, first proposed in 1939. This was a theme that the Institute had tended to ignore in the works that it published during the Weimar era, and during its earliest years in the United States.[22] On the eve of World War II, however, the long-term financial situation of the Institute did not look good, and Horkheimer decided on a scattershot strategy for pursuing funds. At one point, Horkheimer was shepherding proposals for nine projects simultaneously. Horkheimer's willingness to propose a project on anti-Semitism was linked to Horkheimer's sense that Jewish organizations in the United States would be most likely to provide material support to the Institute if they could be convinced that the Institute was conducting research on a topic close to their hearts. The anti-Semitism project, however, was not merely a gambit, but rather a proposed undertaking that Horkheimer explicitly declared he was particularly invested in: "I put more libido into" this proposal, he wrote to a friend, "than in others."[23]

Was it Adorno who proposed that the Institute pursue funds for a project on anti-Semitism? The record does not permit us to answer this question definitively. It may be relevant, however, that when, in the spring of 1939, Horkheimer first began to distribute copies of the project proposal, he characterized it as having been drafted by Adorno and himself.[24] It may also be relevant that it was Adorno who was charged with drafting a second version of the project proposal—a task that he completed with the aid of his wife, at the end of July, 1940.[25]

Days after completing this draft, Adorno described the proposed project to a prominent American who was sympathetic to the Institute as "concerned with the problem of anti-Semitism, especially with such questions as the historic origin of anti-Semitism and the influence of economic developments on anti-semitic psychology and with the different types of present day Anti-Semitism." "We intend" Adorno continued "to include an extensive experimental psychological section. At present we are trying to supplement this project with research into antisemitic theory and practice in National Socialist Germany, as well as its international repercussions."[26]

In a draft of this letter, Adorno had noted that Horkheimer's essay "The Jews and Europe" would provide the theoretical foundation for the proposed project.[27] The fact that no such passage appears in the letter as ultimately sent may well suggest that Adorno was moving beyond the framework of Horkheimer's piece by the summer of 1940.

Not all the members of the Institute of Social Research agreed with Adorno's approach. Franz Neumann—who was explicitly asked by Horkheimer to comment on Adorno's ideas[28]—wrote to Adorno, for example, that he read Adorno as indicating that anti-Semitism was necessary to an understanding of Nazism, but that in fact National Socialism could be presented without allocating a central role to the Jewish problem. anti-Semitism, Neumann asserted, no longer played a major role in the foreign policies of the Nazi regime. Moreover: "I consider the contention that what happens to the Jews actually happens to everyone to be fundamentally wrong"[29] Neumann also underscored in his commentary on Adorno's proposal that members of the Institute ought to be careful, even in drafting project proposals, not to abandon "our theoretical standpoint," and that Adorno's wording, which, Neumann believed, had been adopted for tactical reasons, overestimated the significance of anti-Semitism.[30]

But Adorno's wording was not merely the result of a tactical maneuver, and revealed a genuine distinction between his analysis and that of Neumann. Adorno devoted a considerable amount of thought to the "Jewish question" in the summer of 1940, after he had prepared the second draft of the Institute's project proposal. In a letter to Horkheimer written on August 5, 1940 Adorno writes:

> I am beginning to feel, particularly under the influence of the latest news from Germany, that I cannot stop thinking about the fate of the Jews any more. It often seems to me that everything that we used to see from the point of view of the proletariat has been concentrated today with frightful force upon the Jews. No matter what happens to the project, I ask myself whether we should not say what we want to say in connection with the Jews, who are now at the opposite pole to the concentration of power.[31]

"These lines" Rolf Tiedemann has correctly stressed "provide us with a key to Adorno's thinking from 1940 on."[32]

In the course of a first rate analysis of the construction of anti-Semitism in *Dialectic of Enlightenment* Anson Rabinbach speculates that it was the death of Walter Benjamin, and the arrival, in June of 1941, of Benjamin's testament, "On the Concept of History," which provided Horkheimer and Adorno with a " 'guiding star' around which the constellation of themes—the fate of the exile, the fate of the Jews, and the catastrophe of civilization—that ultimately make up 'Dialectic of Enlightenment' could be organized."[33] It is manifestly the case that Benjamin displayed interest in Judaism over a period of decades, and that this interest (encouraged by Gershom Scholem) had a dramatic impact on his thought and writings.[34] It also universally acknowledged that Benjamin's thought had a profound effect on Adorno.

In this sense, it may well be true that Benjamin, and worry as to Benjamin's fate, increased Adorno's concern with anti-Semitism. However Adorno's letter to Horkheimer of August 5, 1940 demonstrates that Adorno suggested a theoretical project that would revolve around anti-Semitism before the suicide of Adorno's old friend, and before Adorno's reading of Benjamin's last major piece.

Moreover, Rabinbach mistakenly attributes to Horkheimer a crucial letter, dating from October 1941, which discusses the book that Horkheimer and Adorno were planning to cowrite, and thus contributes to a misunderstanding as to why Horkheimer became so keen to write a serious, theoretical, piece on anti-Semitism.[35] It was Adorno—not Horkheimer—who writes: "How would it be if the book [. . .] were to crystallize around anti-Semitism? This would bring with it the concretization and limitation which we have been looking for. It would also be possible for the topic to motivate most of the Institute's associates [. . .]."[36] Adorno's continuing desire to write on anti-Semitism was sparked not by Benjamin's death per se, but by a deeper sense that, as he put it in this letter, anti-Semitism had become the contemporary world's "central injustice," and that it was the task of Critical Theory to "attend to the world where it shows its face at its most gruesome."[37]

Horkheimer found Adorno's reasoning compelling. For this reason (among others) Horkheimer became convinced that he should devote sustained attention to anti-Semitism. He also encouraged the Institute's associates to devote themselves to projects intended to explain the causes of anti-Semitism, its significance, and techniques that could combat it—subjects to which the Institute in exile dedicated years of study, thousands of hours of work, and a considerable sum of money.

Horkheimer's interest in having the Institute write about anti-Semitism, to be sure, was fed by a number of streams.[38] Horkheimer came to believe that work on anti-Semitism would enable the ideas of the Institute to attract "broader interest," that it would provide "an opportunity to develop some of those ideas in a more concrete material" and that it would allow the Institute to present itself as expert "in particular social problems."[39] Horkheimer's Jewish family background, which led him to identify himself as Jewish when speaking to American Jewish audiences, sensitized him to the general issue of Jew hatred. The anti-Semitism encountered by members of Horkheimer's circle in the United States probably made Horkheimer more aware than he had previously been of how pervasive a problem was posed by this phenomenon. Leo Lowenthal later noted that it was not until he and his colleagues had come to the United States that they "suddenly discovered that something like a real everyday anti-Semitism did exist here and that as a Jew one couldn't freely take part in all social spheres. That hotels and clubs, even whole professions, were simply closed to Jews—that didn't yet exist in Germany to such an extent" in the formative years of the members of the Institute.[40] Last not least: the Institute's pressing financial needs—mentioned above—certainly played a major role in Horkheimer's decision to have his associates seek large-scale grants for a series of projects related to anti-Semitism.[41]

However, it was not, in the last analysis, the need for funds that best explains why Horkheimer chose to write on anti-Semitism himself, but rather Adorno's premonition of the destruction of European Jewry, Adorno's repeated suggestions that he and Horkheimer write on anti-Semitism, and a growing sense on the part of both

Horkheimer and Adorno of the significance of continuing Jew hatred. By 1941, Horkheimer—unlike, it can be safely assumed, Neumann—had come to believe, as he wrote in a letter to Harold Laski, that "society itself can be properly understood only through antiSemitism."[42]

It ought to be noted that not only Neumann but also certain other members of the Institute were not immediately convinced. Lowenthal was inclined, during this period, to be heavily influenced by views expressed by the Institute's director. However, as late as the fall of 1942, Lowenthal (who did not maintain his active involvement with Jewish institutions after he became a full time member of the Institute in 1930) frankly admitted to Horkheimer "so far I have not developed a genuine affinity towards the problem itself"[43] and implored Horkheimer to explain to him "the esoteric significance of the whole enterprise."[44] Horkheimer replies by assuring Lowenthal that he would reveal the project's importance in due course, and also notes "These days are days of sadness. The extermination of the Jewish people has reached dimensions greater than at any time in history. I think that the night after these events will be very long and may devour humanity."[45] Members of the Institute who distrusted the development of Critical Theory beyond Marxism had far greater misgivings than those expressed in Lowenthal's letter. Lowenthal reported in 1943, for example, that Henryk Grossmann's "opinions about our research project and about the political situation are as stupid as they always were."[46]

But Adorno, Horkheimer, and Pollock had become committed to the attempt to explain anti-Semitic behavior, and Lowenthal was willing to follow Horkheimer's lead. Adorno underscored as early as September 1940[47] that a theoretical explanation of Jew-hatred would have to depend on a primeval history, and that this history could not be grounded simply in psychological factors, as Freud had attempted, but rather had to be rooted in archaic social movements. In a very preliminary attempt to present to Horkheimer some thoughts as to the origins of anti-Semitism and the reasons for its persistence Adorno writes:

> The survival of nomadism among the Jews might provide not only an explanation for the nature of the Jew himself, but even more an explanation for anti-Semitism. The abandonment of nomadism was apparently one of the most difficult sacrifices demanded in human history. The Western concept of work, and all of the instinctual repression it involves, may coincide exactly with the development of settled habitation. The image of the Jews is one of a condition of humanity in which work is unknown, and all of the later attacks on the parasitic, miserly character of the Jews are mere rationalizations. The Jews are the ones who have not allowed themselves to be "civilized" and subjected to the priority of work. This has not been forgiven them, and that is why they are a bone of contention in class society. They have not allowed themselves, one might say, to be driven out of Paradise, or at least only reluctantly. In addition, the description that Moses gives of the land flowing with milk and honey is a description of Paradise. This holding firm to the most ancient image of happiness is the Jewish utopia. It does not matter whether the nomadic condition was in fact a happy one or not. [. . .] But the more the world of settled habitation—a world of work—produced repression, the more the earlier condition must have seemed to be a form of happiness which could not be permitted, the very idea of which must be banned. This ban is the origin of anti-Semitism, the expulsions of the Jews, and the attempt to complete or imitate the expulsion from Paradise.[48]

Ideas suggested by Adorno in his memo of September 1940 planted the seeds for the approach taken by Horkheimer, Adorno, and Lowenthal in "Elements of Antisemitism," Critical Theory's most famous attempt to provide an explanation for hatred of the Jews. I will not rehash the points made by Horkheimer and Adorno in "Elements," which has been exhaustively analyzed on numerous occasions.[49] Suffice it to say that the authors explain that "[t]he discussion [. . .] of 'Elements of Anti-semitism' deals with the reversion of enlightened civilization to barbarism." They also saw this piece, ultimately published as the final chapter of *Dialectic of Enlightenment*, as "directly related to empirical research by the Institute of Social Research" (that is to say the projects carried out by the Institute with the aid and support of the American Jewish Committee and of the Jewish Labor Committee).[50] Indeed, when Adorno was preparing a questionnaire to be used in conjunction with one of these empirical projects he told Horkheimer that he had distilled a number of the questions from the "Elements of Antisemitism" through a kind of translation process.[51]

In the period following the drafting of the first theses of the "Elements of Antisemitism" in 1943, Horkheimer attempted to explain anti-Semitism and Jewish history by pointing to both psychological and historical factors. To be sure, Horkheimer, when addressing the outside world, would, by the end of 1943, write that "despite the tremendous importance of economic and social tendencies, antiSemitism is fundamentally a psychological phenomenon."[52] When writing confidentially to Pollock during this period, on the other hand, Horkheimer stresses that "[o]ur real ideas on AntiSemitism attribute an infinitely greater part to economic and social factors than an isolated glimpse upon our work could suggest."[53]

Horkheimer also repeatedly argued during this period that anti-Semitism was a spearhead, that is, that it was "not alone a menace to the Jews, but a symptom of the crisis facing democratic civilization."[54] "Hatred of the Jew" Horkheimer insisted "is hatred of democracy [. . .]."[55] Thus, "an organized investigation into the psychology of antisemitism is in no way a matter concerning Jews exclusively, but it is of the greatest importance for all who are interested in the fate of democratic civilization."[56]

In a recent commentary on the *Dialectic of Enlightenment* David Seymour criticizes Horkheimer and Adorno for failing to examine Jews. According to Seymour, the "Elements of Antisemitism" explains anti-Semitism as "a form of 'projection' which has little or nothing to do with Jews themselves and is attached to Jews regardless of their own actions and agency."[57] A fuller examination of the ideas of Horkheimer and Adorno, relying not only on the *Dialectic* but also on unpublished writings by them on hatred of the Jews, suggests that the Critical Theorists recognized the need to examine not only anti-Semites, but also the objects of anti-Semitism. Horkheimer, for example, wrote to Pollock in May of 1943 that "it is not only imperative to enter into a thorough study of the antisemitic reactions, but of Jewish psychology as well."[58] Horkheimer prepared a few thoughts on the psychology of Jews for a speech he gave in 1944—but seems to have left these thoughts out of the speech he ultimately delivered.[59] However, it is also true that the point that Horkheimer and Adorno were making in their published writings was that it was not so much the Jews themselves as it was the image of the Jews that was significant in explanations of contemporary anti-Semitism: "No matter what the Jews as such may be like" reads a revealing passage in *Dialectic of Enlightenment,*

> their image, as that of the defeated people, has the features to which totalitarian domination must be completely hostile: happiness without power, wages without work, a home without frontiers, religion without myth. These characteristics are hated by the rulers because the ruled secretly long to possess them. The rulers are only safe as long as the people they rule turn their long-for goals into hated forms of evil.[60]

The actualization of Adorno's foreboding prediction of 1938 led Adorno to underscore, as early as the autumn of 1944, that civilization had been profoundly altered:

> The idea that after this war life will continue "normally" or even that culture might be "rebuilt"—as if the rebuilding of culture were not already its negation is idiotic. Millions of Jews have been murdered, and this is to be seen as an interlude and not the catastrophe itself. What more is this culture waiting for? And even if countless people still have time to wait, is it conceivable that what happened in Europe will have no consequences, that the quantity of victims will not be transformed into a new quality of society at large, barbarism?[61]

From Adorno's perspective, anti-Semitism had changed the whole world, let alone the world of the "Bildungsbürger" in which the term "culture" had been raised as the banner of righteousness.

Anti-Semitism had also changed Adorno's own relationship to his Jewish roots. Though not self-identified as a Jew in the pre-Holocaust years, Adorno seems to have come to see himself as a Jewish Holocaust survivor as the enormity of events in Europe became apparent (and was seen as a Jewish figure by many postwar Germans). *Minima Moralia,* written during and immediately after World War II, concludes with a deeply Jewish passage that could easily have been written by Benjamin.[62] Later work, such as *Negative Dialectics,* seems to confirm this shift in self-perception.[63]

Even during the course of the 1940s, Adorno was worried that readers might believe, inaccurately, that the Institute's explanations of Antisemitism were grounded primarily in psychology or philosophy. In order to overcome such a misunderstanding, Adorno told his colleagues in August of 1944 that "We should stress that the so-called psychology of anti-semitism does not operate in an empty space, but is essentially and intrinsically related to social situations [. . .]."[64] In 1949, in the months leading up to Adorno's return to Frankfurt, he continued to attempt to whip the Institute's massive study of anti-Semitism among American workers into publishable shape, and argued at one point that there ought to be a theoretical chapter added to the end of the labor study, which would explain the situation confronting the American working class by reference to the contradiction between its objective position in the process of production and its subjective consciousness of itself.[65] But, despite years of work, the Labor Study never did appear in print. And thus the misunderstanding that Adorno feared has persisted.

If one compares Horkheimer's essay on the Jews and Europe with relevant writings written over the course of the next few years, I conclude, it becomes manifest that

Horkheimer's thinking shifted considerably. In the period just before the War began, Horkheimer still placed particular stress on the primacy of economics in explaining the phenomenon of anti-Semitism. In pieces written during the War by Horkheimer and by Adorno dealing with anti-Semitism, economics is merely one part of a multi-faceted explication. The shift in Horkheimer's perspective, I have argued, was part and parcel of a larger shift that deeply affected Critical Theory as a whole, and is best explained by the influence of Adorno, and by reflection on the destruction of the Jews in progress.

However this shift did not signify a total rejection of previously held positions. Horkheimer, Adorno, and those Critical Theorists closest to them, intended to supplement works like *The Authoritarian Personality* with writings in which anti-Semitism would be explained as intimately intertwined with economic and social conditions. The objections by Neumann and those members of the Institute who were ideologically close to him could not block the line of Horkheimer and Adorno's development, but they did represent something of a Marxist conscience and a reminder of unfinished business. Moreover: these objections seem to have slowed Horkheimer's drift away from his initial position. The published pieces by Horkheimer and Adorno, which grapple with anti-Semitism and which date from the exile period, are potentially misleading if read in a vacuum insofar as they suggest that Critical Theory had given up altogether on its Marxist roots. But "an aversion to closed philosophical systems" as Martin Jay first told us "lies at the very heart of Critical Theory."[66] Thus the fact that the published writings of the Critical Theorists on anti-Semitism and the image of the Jew have remained "philosophical fragments" strikes me as ultimately not unfitting.

Notes

1. Douglas Kellner, "The Frankfurt School Revisited: A Critique of Martin Jay's *The Dialectical Imagination*" (1973), in *The Frankfurt School. Critical Assessments*, ed. Jay Bernstein (London, New York: Routledge, 1994), 41–62, 44, 50.
2. Helmut Dubiel, *Theory and Politics. Studies in the Development of Critical Theory* (1978), trans. Benjamin Gregg (Cambridge, Mass.: MIT Press, 1985), 105.
3. Martin Jay, "The Jews and the Frankfurt School: Critical Theory's Analysis of Antisemitism" (1979), in *Permanent Exiles. Essays on the Intellectual Migration from Germany to America* (New York: Columbia University Press, 1986), 90–100, 93.
4. Max Horkheimer, "The Jews and Europe" (1939), in Stephen Eric Bronner and Douglas MacKay Kellner (ed.), *Critical Theory and Society. A Reader* (New York, London: Routledge, 1989), 77–94, 78, 92.
5. Cf. Theodor W. Adorno to Max Horkheimer, August 10, 1939, Max-Horkheimer-Archiv, Stadt- und Universitätsbibliothek, Frankfurt-am-Main, VI 1A 47-47B, in which Adorno comments on a draft of Horkheimer's article.
6. Theodor W. Adorno to Walter Benjamin, July 15, 1939, in Theodor W. Adorno and Walter Benjamin, *The Complete Correspondence 1928–1940*, ed. Henri Lonitz, trans. Nicholas Walker (Cambridge/Mass.: Harvard University Press, 1999), 313 (hereafter cited as *The Complete Correspondence 1928–1940*).
7. Theodor W. Adorno to Max Horkheimer, February 15, 1938, in Horkheimer, *Gesammelte Schriften*, XVI, 392; Gunzelin Schmid Noerr, *Gesten aus Begriffen. Konstellationen der Kritischen Theorie* (Frankfurt: Fischer, 1997), 133.

8. Marion A. Kaplan, *Between Dignity and Despair. Jewish Life in Nazi Germany* (New York, Oxford: Oxford University Press, 1998), 5.

9. Rolf Wiggershaus, *The Frankfurt School: Its History, Theories, and Political Significance* (1986), trans. Michael Robertson (Cambridge/Mass: The MIT Press, 1994), 66–67 (hereafter cited as *The Frankfurt School*).

10. Theodor W. Adorno to Ernst Krenek, October 7, 1934, in: Theodor W. Adorno and Ernst Krenek, *Briefwechsel* (Frankfurt: Suhrkamp, 1974), 46 (hereafter cited as *Briefwechsel*).

11. Leo Lowenthal, *An Unmastered Past. The Autobiographical Reflections of Leo Lowenthal*, ed. Martin Jay. Berkeley (Los Angeles, London: University of California Press, 1987), 205 (hereafter cited as *An Unmastered Past*).

12. Peter von Haselberg, "Wiesengrund-Adorno," in Heinz-Ludwig Arnold (ed.), *Theodor W. Adorno*, second edn. (Munich: Edition Text + Kritik, 1983), 7–21, 12.

13. Adorno, *The Curious Realist*, 161.

14. Theodor W. Adorno to Alban Berg, September 8, 1933, in Theodor W. Adorno and Alban Berg, *Briefwechsel 1925–1935*, ed. Henri Lonitz, Briefe und Briefwechsel II (Frankfurt: Suhrkamp, 1997), 275 (hereafter cited as *Briefwechsel 1925–1935*). Theodor W. Adorno to Alban Berg, November 13, 1933, in Adorno and Berg, *Briefwechsel 1925–1935*, 279.

15. Wiggershaus, *The Frankfurt School*, 157.

16. Alban Berg to Theodor W. Adorno, November 18, 1933, in Adorno and Berg, *Briefwechsel 1925–1935*, 285.

17. Theodor W. Adorno to Alban Berg, November 28, 1933, in Adorno and Berg, *Briefwechsel 1925–1935*, 286–287.

18. Steven S. Schwarzschild, "Adorno and Schoenberg as Jews. Between Kant and Hegel," *Leo Baeck Institute Year Book* XXXV (1990), 443–478, 455. Wilcock, *Negative Identity*, 178–182.

19. Theodor W. Adorno to Ernst Krenek, October 7, 1934, in Adorno and Krenek, *Briefwechsel*, 43–44.

20. Walter Benjamin to Theodor W. Adorno, May 7, 1940, in Adorno and Benjamin, *The Complete Correspondence 1928–1940*, 330.

21. Theodor W. Adorno to Max Horkheimer, January 14, 1940, in Horkheimer, *Gesammelte Schriften*, XVI, 696.

22. Jack Jacobs, "A Most Remarkable 'Jewish Sect'? Jewish Identity and the Institute of Social Research in the Years of the Weimar Republic," *Archiv für Sozialgeschichte* 37 (1997), 73–92, here: 73–74, 76–78.

23. Max Horkheimer to Katharina von Hirsch, July 14, 1939, in Horkheimer, *Gesammelte Schriften*, XVI, 615.

24. Max Horkheimer to Juliette Favez, April 25, 1939, in Horkheimer, *Gesammelte Schriften*, XVI, 472. The initial project proposal seems to have been completed in April of 1939, at which time it was sent by Horkheimer to Ernst Simmel [Max Horkheimer to Ernst Simmel, April 21, 1939, in Horkheimer, *Gesammelte Schriften*, XVI, 585–586].

25. Theodor W. Adorno to Max Horkheimer, July 29, 1940, in Horkheimer, *Gesammelte Schriften*, XVI, 734.

26. Theodor W. Adorno to Charles E. Merriam, July 30, 1940, in Horkheimer, *Gesammelte Schriften*, XVI, 743.

27. Horkheimer, *Gesammelte Schriften*, XVI, 746.

28. Horkheimer to Franz Neumann, July 10, 1940, Max-Horkheimer-Archiv, Stadt- und Universitätsbibliothek, Frankfurt-am-Main, VI, 30, 117.

29. Franz Neumann to Theodor W. Adorno, August 14, 1940, Max-Horkheimer-Archiv, Stadt- und Universitätsbibliothek, Frankfurt-am-Main, VI 1A 22.

30. A project proposal all but certainly based on the proposal prepared by Adorno was published as "Research Project on Anti-Semitism," in *Studies in Philosophy and Social Science*, IX, 1941, 124–143.

31. Theodor W. Adorno to Max Horkheimer, August 5, 1940, in Wiggershaus, *The Frankfurt School*, 275.
32. Rolf Tiedemann, "Introduction: 'Not the First Philosophy, but a Last One': Notes on Adorno's Thought," in Theodor W. Adorno, *Can One Live after Auschwitz? A Philosophical Reader*, ed. Rolf Tiedemann, trans. Rodney Livingstone and others (Stanford, California: Stanford University Press, 2003), xi–xxvii, xix.
33. Anson Rabinbach, "The Cunning of Unreason: Mimesis and the Construction of Antisemitism in Horkheimer and Adorno's *Dialectic of Enlightenment*," in *In the Shadow of Catastrophe. German Intellectuals between Apocalypse and Enlightenment* (Berkeley, Los Angeles, London: University of California Press, 1997), 166–198, 174 (hereafter cited as *The Cunning of Unreason*).
34. Gershom Scholem, *On Jews and Judaism in Crisis. Selected Essays*, ed. Werner J. Dannhauser (New York: Schocken Books, 1976), 190–197.
35. Anson Rabinbach, " 'Why Were the Jews Sacrificed?' The Place of Antisemitism in Adorno and Horkheimer's *Dialectic of Enlightenment*," in Nigel Gibson and Andrew Rubin (eds.), *Adorno. A Critical Reader* (Malden, Massachusetts: Blackwell Publishers Inc., 2002), 132–149, 137.
36. Theodor W. Adorno to Max Horkheimer, October 2, 1941, in Wiggershaus, *The Frankfurt School*, 309. Cf. Max-Horkheimer-Archiv, Stadt- und Universitätsbibliothek, Frankfurt-am-Main, VI, 32, 29.
37. Ibid.
38. Max Horkheimer to Friedrich Pollock, November 23, 1941, Max-Horkheimer-Archiv, Stadt- und Universitätsbibliothek, Frankfurt-am-Main, VI, 32, 29.
39. Max Horkheimer to Herbert Marcuse, April 3, 1943, Max-Horkheimer-Archiv, Stadt- und Universitätsbibliothek, Frankfurt-am-Main, VI, 27A, 21.
40. Lowenthal, *An Unmastered Past*, 30–31.
41. Max Horkheimer to Friedrich Pollock, March 20, 1943, Max-Horkheimer-Archiv, Stadt- und Universitätsbibliothek, Frankfurt-am-Main, VI, 33.
42. Wiggershaus, *The Frankfurt School*, 690.
43. Leo Lowenthal to Max Horkheimer, October 27, 1942, Max-Horkheimer-Archiv, Stadt- und Universitätsbibliothek, Frankfurt-am-Main, VI, 15, 315.
44. Leo Lowenthal to Max Horkheimer, October 29, 1942, Max-Horkheimer-Archiv, Stadt- und Universitätsbibliothek, Frankfurt-am-Main VI, 15, 306.
45. Max Horkheimer to Leo Lowenthal, November 27, 1942, in Horkheimer, *Gesammelte Schriften*, XVII, 384.
46. Leo Lowenthal to Max Horkheimer, August 20, 1943, Leo Lowenthal Papers, Harvard, bMS Ger 185 [78], folder 23.
47. In a memo sent with Theodor W. Adorno to Max Horkheimer, September 18, 1940, in Horkheimer, *Gesammelte Schriften*, XVI, 760–761.
48. Wiggershaus, *The Frankfurt School*, 276–277.
49. See, for example, the recent discussion in Brett R. Wheeler, "Antisemitism as Distorted Politics: Adorno on the Public Sphere," *Jewish Social Studies* 7, 2 (2001): 114–148.
50. Horkheimer and Adorno, *Dialectic of Enlightenment*, xix.
51. Theodor W. Adorno to Max Horkheimer, November 9, 1944, Max-Horkheimer-Archiv, Stadt- und Universitätsbibliothek, Frankfurt-am-Main, VI 1B 180. For additional information on the project carried out with the financial support of the Jewish Labor Committee see Jack Jacobs, "1939: Max Horkheimer's 'Die Juden und Europa,' " in Sander L. Gilman and Jack Zipes (eds.), *Yale Companion to Jewish Writing and Thought in German Culture 1096–1996* (New Haven and London: Yale University Press, 1997), 571–576, 574–575.
52. Max Horkheimer, "Notes on Horkheimer's Remarks," in Horkheimer, *Gesammelte Schriften*, XVII, 521.

53. Max Horkheimer to Friedrich Pollock, July 3, 1943, Max-Horkheimer-Archiv, Stadt- und Universitätsbibliothek, Frankfurt-am-Main, VI, 33, 419.
54. Max Horkheimer, "Notes on Horkheimer's Remarks," in Horkheimer, *Gesammelte Schriften*, XVII, 526.
55. Speech by Max Horkheimer delivered at Temple Israel, April 30, 1943, Max-Horkheimer-Archiv, Stadt- und Universitätsbibliothek, Frankfurt-am-Main, IX, 44A. Cf. Horkheimer, *Gesammelte Schriften*, XII, 172.
56. Max Horkheimer, "Notes on Horkheimer's Remarks," in Horkheimer, *Gesammelte Schriften*, XVII, 521–522.
57. Seymour, 2000, 300.
58. Max Horkheimer to Friedrich Pollock, May 19, 1943, Max-Horkheimer-Archiv, Stadt- und Universitätsbibliothek, Frankfurt-am-Main, VI, 33, 506.
59. Max Horkheimer, "Antisemitism as a Social Phenomenon," June 17, 1944, Max-Horkheimer-Archiv, Stadt- und Universitätsbibliothek, Frankfurt-am-Main IX, 46, 1A, 10. There is a handwritten notation on this typescript, "close here," immediately preceding Horkheimer's comments on the psychology of the Jews.
60. Horkheimer and Adorno, *Dialectic of Enlightenment*, 199.
61. Theodor W. Adorno, *Minima Moralia. Reflections from a Damaged Life* (1951), trans. E. F. N. Jephcott (London: Verso, 1974), 55 (hereafter cited as *Minima Moralia*).
62. "The only philosophy which can be responsibly practised in face of despair is the attempt to contemplate all things as they would present themselves from the standpoint of redemption. Knowledge has no light but that shed on the world by redemption: all else is reconstruction, mere technique. Perspectives must be fashioned that displace and estrange the world, reveal it to be, with its rifts and crevices, as indigent and distorted as it will appear one day in the messianic light." (Adorno, *Minima Moralia*, 247).
63. Theodor W. Adorno, *Negative Dialectics* (1966), trans. E. B. Ashton (New York: Continuum, 1973), 363.
64. "Re: Editing of Report to the A. J. C. (Dr. Klein)," August 1944, Max-Horkheimer-Archiv, Stadt- und Universitätsbibliothek, Frankfurt-am-Main, VI 1B 218.
65. Theodor W. Adorno to Leo Lowenthal, May 4, 1949, Leo Lowenthal Papers, Harvard, bMS Ger 185 [8], folder 7.
66. Martin Jay, *The Dialectical Imagination. A History of the Frankfurt School and the Institute of Social Research, 1923–1950* (Boston, Toronto: Little, Brown and Company, 1973), 41.

CHAPTER TWELVE

NOT-SUCH-ODD COUPLES: PAUL LAZARSFELD AND THE HORKHEIMER CIRCLE ON MORNINGSIDE HEIGHTS

Thomas Wheatland

We have inherited an almost mythic image of the relations between members of the Horkheimer Circle and Paul Lazarsfeld. Although both parties exemplified the kind of professional cooperation that exile at times necessitated, traditional accounts depict these two sociological entities as, at most, strange bedfellows. Especially among the scholars and partisans of Critical Theory, Lazarsfeld typified the empirical while the Horkheimer Circle embodied the theoretical (or cultural, or critical) approach to social science. Lazarfeld represented the culmination of applied mathematics in a corporate model of sociology, and the Horkheimer Circle remained the heirs of *Bildung* in their aspirations to fashion a comprehensive theory of contemporary society particularly oriented toward the transformation of the existing social order. Ultimately this awkward relationship has been personified in the antipathies between Paul Lazarsfeld and Theodor W. Adorno that are given prominence in so much of the literature regarding these two figures.[1]

In reality, Lazarsfeld worked with nearly all members of the Horkheimer Circle, and these collaborations began in the early 1930s while Lazarsfeld was still working in Vienna and the Horkheimer Circle still resided in Frankfurt. During the first years of both parties' overlapping sojourn in America, moreover, Lazarsfeld assisted Erich Fromm in the development and execution of the Horkheimer Circle's social research projects on authority, unemployment, and the family. Nevertheless, the story *of* Lazarsfeld's brief unhappy collaboration with Adorno overshadows all of Lazarsfeld's other encounters with the Institute for Social Research. The *caricatured view* of this "odd couple" is no doubt partly due to the fact that both Adorno's and Lazarsfeld's *vivid* retrospective accounts of their relationship appeared just when the "Frankfurt School" mythology was at its height.[2] Both elected to comment feelingly and at length on their brief and unsuccessful collaboration on the Princeton Radio Research Project. For Adorno, the experience with Lazarsfeld embodied his struggle against American research methods and positivism. At the same time, Lazarsfeld recalled his frustrations with Adorno and the exotic and stubborn "foreignness" of his thought.

These reminiscences quickly moved beyond the realm of the private with their publication in the charged atmosphere of the late 1960s. While "radical sociologists" were on the rise and challenging the successful marriage of Functionalism and empiricism that molded postwar U.S. sociology, Adorno and Lazarsfeld seemed to embody the two sides of the emergent battle lines. In the minds of many, Lazarsfeld's name was synonymous with several of the major developments of postwar sociology. His work with Robert Merton provided a fruitful path not only fortifying Functionalism, but also this partnership signaled how empiricism and theory might successfully come together. More importantly Lazarsfeld's Bureau for Applied Social Research became a model for sociological research centers. Adorno, on the other hand, seemed to epitomize the antithesis of a figure like Lazarsfeld. Fresh on the heels of the Postivists' Dispute,[3] Adorno seemed to reject many of the same assumptions in postwar sociology as the partisans of the New Left. To many sociological radicals, Adorno represented the essence of the critical approach to society and initiated many on the path into Germany's intellectual heritage—a journey into German Romanticism and the tradition of *Bildung* that was fueled by a desire to add nourishment to the disintegrating student movement. Thus, the interest in figures like Marcuse and Adorno quickly led to new scholarly enthusiasms for Lukàcs, Korsch, and the entire Hegelian tradition.

Lazarsfeld and Adorno, however, were not just symbolic pawns used by others to promote professional and intellectual agendas. Both Lazarsfeld and the "Frankfurt School" were interested in exaggerating the same divide. By de-emphasizing their relationship with Lazarsfeld, members of the Horkheimer Circle were able to construct a collective image of geniuses stubbornly isolated in an alien land. Separating themselves from an émigré success story like Lazarsfeld enabled the "Frankfurt School" to exaggerate the marginality of the Institute for Social Research in the United States. In addition to emphasizing the importance of exile in their collective thought, this institutional self-image also helped to cloak many of the Horkheimer Circle's failures in the United States. Furthermore, the juxtaposition with a figure like Lazarsfeld helped to highlight the methodological, institutional, and practical differences between the Horkheimer Circle's concept of social science and the predominant paradigm in both United States and European sociology. Did they overdo their caricatures of Lazarsfeld and others that they accused of the sin of Positivism? Of course they did. But at the same time, Lazarsfeld did not entirely attempt to bridge this rhetorical divide set in motion by Adorno and others. Lazarsfeld was content to remain at a distance from the "Frankfurt School" phenomenon. The Horkheimer Circle had many public faces in the wake of World War II. On the one hand, the *Institut für Sozialforschung* was a major West German center for the promotion of American social research methods. Not surprisingly, Lazarsfeld actively supported such initiatives, and the Frankfurt Institute was a postwar affiliate of the Bureau for Applied Research.[4] Consequently, it should perhaps be of little surprise that Lazarsfeld saw Adorno's postwar attacks on empirical sociology as a personal attack on the Horkheimer Circle's American allies. It is little wonder that Lazarsfeld and many of his associates were all too happy to minimize the extent of the émigré cooperation that took place between the forerunners of the Bureau for Applied Social Research and the "Frankfurt School."

My research focusing on the émigré experiences and receptions of the Horkheimer Circle in the United States has uncovered a relationship between Lazarsfeld and the Institute for Social Research that was much closer and also far more complex than the existing literature suggests.[5] Lazarsfeld was involved in most of the Institute's socio-logical investigations throughout the late 1930s; he was instrumental in helping the group seek grant support for their research proposals of the early 1940s; he was a con-stant advisor during the final stages of the completion of the *Studies in Prejudice*; and he most likely served as the inspiration and role model for *The Studies in Prejudice* when they were first set in motion by the Horkheimer Circle. While the extent of these relations was moderately surprising, the context for this collaboration was a revelation. The behavior of Lazarsfeld and the Horkheimer Circle is only partly acces-sible to us without a better understanding of the academic and intellectual environ-ments shaping their interactions. Columbia University, which eventually became the host institution for both the Horkheimer Circle and Paul Lazarsfeld, powerfully affected the relationships between the two parties.

A certain degree of mystery surrounds the invitation of Horkheimer's Institute to Morningside Heights.[6] While the motives of the Institute's move to New York have always been clear, Columbia's motives regarding the Institute for Social Research are impenetrably murky. What interests would Columbia's President, Nicholas Murray Butler, and its sociology department have had in an avowedly leftist Institute populated by an overwhelming majority of social philosophers who spoke relatively little English?

In the years leading up to Hitler's rise to power and the ensuing "Intellectual Migration" that commenced in 1933, Columbia University's sociology department faced a serious crisis. Under the leadership of the great Franklin Giddings, Columbia sociology established itself as the premier institution for the study of social science during the late nineteenth century. Although Columbia (under Giddings) success-fully helped to model the social sciences on the empirical methods of the natural sciences, Robert Park more quickly mastered the new approach to the discipline. The University of Chicago used the generous support of the Rockefellers to pioneer the subfield of community studies and surpassed Columbia by the late 1920s. Eager to regain their stature, Columbia's sociologists agreed that a quantitative research insti-tute gave their department the best opportunity for revitalization. In 1928, Columbia's President, Nicholas Murray Butler, toyed with the idea of eliminating sociology at Columbia. Suspicious of its political orientation and reformist aims, Butler probably would have followed through on his plan were it not for the intervention of John Dewey. In the wake of this close call, Columbia's sociologists met frequently to discuss the future. Best expressing their views in a memorandum of 1928, they recorded the following assessment of their discipline:

> But the rise of the quantitative method means, from the administrative and budgetary point of view of the universities, revolutionizing the concept of research. The quantitative method is expensive, and must, distinctly, be seen on this plane. It is a big idea and needs to be treated in a big way. Just as the laboratory in the physical sciences was a big idea and meant large appropriations for equipment, materials, care, assistants, etc. If the universities do not meet this need in a proper way, they must inevitably become insignificant in a func-tion which is properly their own. The achievement and prestige will pass to the government [. . .] or to private industries such as banks and large corporations.[7]

The department proposed to create a quantitative sociological institute on Morningside Heights. Columbia's sociologists estimated that a minimum budget of $50,000 would be necessary obtain the needed calculators, travel budgets, and staff salaries. The unforeseen obstacle that altered the implementation of their plan was the Great Depression of 1929. In the wake of the Crash, the thoughts of buying new equipment and adding new faculty members to the department became pipe dreams. Sociology at Columbia actually shrank in the next decade. The department's leaders throughout the 1930s and 1940s, Robert MacIver and Robert Lynd, accepted these new fiscal realities, but they did not abandon the earlier plans for renewal. They simply became more creative and flexible in thinking about how to bring large-scale, quantitative research to Morningside Heights.

Although their rivalry and mutual contempt are familiar to many, Lynd and MacIver shared at least one common goal—to renew the reputation of Columbia's sociology department institutionally and intellectually. In fact many of their most bitter disagreements related to the best strategies for achieving this ambition. Specifically, they clashed over what types of institutions and individuals were needed to revitalize the department. For Lynd, social science was useless unless its ultimate goal was social reform and transformation—a notion that was similarly articulated and embraced by members of the Horkheimer Circle, as well as many other social scientists that Lynd championed. Where Lynd may have differed with members of the Horkheimer Circle were in his temporary enthusiasm for statistical research and in his views regarding the professionalization of social science. For Robert MacIver, by contrast, the object of sociology was to better understand and theoretically describe society. While MacIver's discomfort with political partisanship in the social sciences was well known, his concern with the rising sociology of minutiae (that he felt was an inevitable companion of empiricism) and his preference for social theorizing made him an occasional supporter of the Horkheimer Circle despite the clearly political implications of their own approach to social theory. When Lynd met Paul Lazarsfeld in 1933, the young Austrian might have seemed an ideal candidate for Columbia's plans. Coming of age among Vienna's Austro-Marxist intellectuals, Lazarsfeld devoted his substantial talents in applied mathematics to the study of social psychology. To help defray the costs connected with his research projects regarding childhood education, career choices, and unemployment, Lazarsfeld created one of the first bureaus for market research. While Lazarsfeld's *Research Bureau for Economic Psychology* assisted Austrian corporations in understanding consumer preferences, Lazarsfeld used these investigations to satisfy his interests in empirical social research and this allowed him to sharpen his skills in various research methodologies. As long as contract work for private industry helped support the many interests of Lazarsfeld's team, he would serve any client, including Frankfurt's *Institut für Sozialforschung*. In recognition for his professional achievements, Lazarsfeld was awarded a Rockefeller fellowship that enabled him to visit the United States in 1933–1934 and to travel to different universities to meet a wide array of U.S. social researchers. The 1934 coup by Austria's conservative Catholic party and the ensuing collapse of democracy in his homeland helped to convince Lazarsfeld to remain in America.

Despite his prestigious Rockefeller grant, Lazarsfeld remained only a future prospect for Lynd during the early 1930s. The obstacles created by the circumstances of his emigration made it impossible for Lazarsfeld to be an immediate solution to the problems facing Columbia sociology. Lazarsfeld's institutional infrastructure remained in Vienna, and there was little short-term hope that the resources would be available to quickly replicate such a self-sustaining operation in the United States. Lynd generously offered his support, and assisted Lazarsfeld's appointment at the University of Newark. Lynd liked and admired Lazarsfeld, and he always supported underdogs. In Lynd's view, Lazarsfeld's background and interests might have seemed a perfect match with the type of research that Lynd sought to promote. Lazarsfeld was empirically rigorous but committed to research that addressed the same kind of social problems that concerned Lynd (with the notable exception of the corporate research). Although his own leftist politics were formed through his training in the Church and through his experiences in "Middletown," Lynd sympathized with many of the convictions that Lazarsfeld forged at home and in Vienna as a student. Furthermore, Lazarsfeld's remarkable talents as a quantitative researcher and experiences guiding highly professionalized research institutions also impressed Lynd. Many of Lazarsfeld's thoughts regarding the purpose, organization, and pedagogy of sociology corresponded closely to Lynd's ideas during this phase of Lynd's career. Ironically in subsequent years and partly as a result of his later work with Lazarsfeld, Lynd grew to become frustrated and troubled by precisely the same professionalism and methods that he had admired years before.

Although we do not normally associate the Horkheimer Circle with quantitative research methods, the Institute for Social Research appeared to Columbia's sociologists in 1933 as the answer to their ambitions. The Institute did not misrepresent itself to Columbia University, but it did emphasize the use of quantitative methods in its research. While Columbia's sociologists almost certainly would have had little interest in the Horkheimer Circle's *Zeitschrift für Sozialforschung*, the projects forming the empirical skeleton for *Studien über Autorität und Familie* would have appeared extremely enticing. Furthermore, the commitment to social reform espoused by the Horkheimer Circle would have also appealed to the older members of the Columbia faculty that still remembered U.S. sociology's relationship to the "*social gospel*," as well as to younger progressives like Robert Lynd. The combination seemed to be reminiscent of Gidding's own approach to sociology, but the institute's framework promised to serve as the spark that Columbia sought. As the text from the constitution for the International Society of Social Research (the umbrella organization controlling the purse strings of the Institute) read,

> The sole object of the corporation shall be to render social service. It will study problems of international welfare, and will endeavor to solve the same by means of social research, the application of scientific knowledge, and by assembling necessary data and other appropriate means.[8]

Erich Fromm and Julian Gumperz, who led the Institute's search for a U.S. home, echoed the approach to social research that Horkheimer had been repeating since his inaugural address as Director of the Institute in 1931.[9] They wrote and spoke of the dialectical mediation of empirical research methods and social philosophy,[10] but it was predominantly the Institute's quantitative social research that caught the attention of Columbia's sociologists. One of the Horkheimer Circle's staunchest supporters at Columbia, Robert Lynd, also could not have ignored the fact that this group of émigrés was interested in a topic remarkably similar to one that he had in mind to study for several years prior to their arrival in New York—the impact of parental unemployment on the modern family.[11] They were further fortunate that Lynd had also been calling for interdisciplinary cooperation in the social sciences that bore resemblance to the Institute for Social Research's self-appraisal and the presentation of itself that was provided to Columbia University. As Lynd explained during a conference of the Social Science Research Council in February of 1931:

> There are different patterns of cooperation in research, some of them amounting only to a kind of pseudo-cooperation. At it simplest, cooperative research may consist in giving the expert a corps of pick-and-shovel junior assistants to speed along his labors. A more complicated pattern involves the association of a group of specialists in a single discipline for mutual inter-stimulation, and in some cases, joint attack upon a common problem. Still a third pattern, the most complicated of all, is the association of men from radically diverse disciplines for joint action upon some problem sprawled across all their fields. The first two of these patters are fairly common and the third as yet largely a matter of occasional experimental forays.[12]

Lynd and the other members of Columbia's sociology department must have been surprised by what the Institute appeared to offer and by what Columbia stood to gain. This Institute for Social Research sought an informal association with their university, and the only carrot dangled by President Butler was the Institute's use of a run-down brownstone at 428 W, 117th Street. The Institute would be expected to remain self-sufficient—its endowment would continue to pay for its personnel and research projects, and it was even responsible for funding the renovation and upkeep of the building loaned by Columbia. For this negligible price, Columbia's sociologists expected to gain a great deal—an institution that was capable of fulfilling the revitalization of sociology at Columbia while simultaneously appealing to the two scholarly factions that had arisen in Columbia's department during the 1930s. The Horkheimer Circle, by clearly sharing its commitments to social reform and social transformation, were sure to appeal to the political progressives rallying around Robert Lynd. Furthermore, the incorporation of statistical methods and analysis in the Institute's large-scale research projects was an additional attraction to the partisans of the quantitative method also supporting Lynd. At the same time, the Horkheimer Circle's theoretical work and cautious attitude toward Positivism were enough to ease any reservations held by Robert MacIver and the social theorists and political scientists allied with him. There is no question that the sociopolitical implication of Critical Theory must have troubled MacIver, but he was able to locate a common theoretical ground with these émigrés despite the fact that the Institute's legacy was German Idealism and his own was the Anglo-Scottish Enlightenment. The Institute for Social Research, thus, was a perfect compromise for Columbia's sociologists while offering the potential for departmental revitalization at no cost to the university.

For those familiar with the Horkheimer Circle's history in America, the course of this relationship was entirely predictable. Columbia's sociologists never realized the extent to which the Institute relied on research associates across Europe who would not be a part of the future projects planned for the United States. Erich Fromm, the architect and manager of the Institute's research endeavors, soon became a recognizable and popular figure among the Faculty of Political Science on Morningside Heights, but the other members of the Horkheimer Circle remained more committed to their German contributions to the *Zeitschrft für Sozialforschung*. In addition to completing the Institute's massive *Studien über Autorität und Familie* and struggling to sift through collected data regarding the German working class, Fromm initiated new research projects that involved Columbia sociologists and their students. One was a study of unemployment and its impact on American families in Newark, and the second examined postadolescent attitudes toward authority among the population of students at nearby Sarah Lawrence College. Despite the administrative role assumed by Fromm and the Institute's financial support for both projects, the Horkheimer Circle did not possess an experienced research staff in exile to execute these projects. Nearly all of the data connected with both endeavors was gathered and analyzed by Columbia sociology professors and their students (in fact Mirra Komarovsky, a junior researcher on Morningside Heights, was so involved with the Newark project that the Institute for Social Research supported her publication of the findings),[13] Paul Lazarsfeld's new social science bureau at the University of Newark, and researchers at Sarah Lawrence (which resulted in another publication which was never credited to the Institute for Social Research).[14] By the time the Institute's endowment was decimated by poor money management and the recession of 1938, the relationship with Fromm became damaged beyond repair. When Fromm ended his association with the Horkheimer Circle, the social research that he directed also ended. Now the "Frankfurt School" was exposed for what it largely was—a modest array of social philosophers practicing a method of Critical Theory that generally could not yet be appreciated by American scholarly audiences. Columbia's sociologists, most of whom liked and admired Fromm, grew to view the Institute with frustration and impatience. The department's gamble, which appeared to be a no-lose situation, had hit a serious snag. Without Fromm, all of the empirical projects vanished from Morningside Heights.

The chief obstacle to Horkheimer's plan for new social research projects that could receive grant support and repair the damaged relationship with Columbia was the group's inexperience with designing, managing, and completing empirical social research. Despite this obstacle and their ignorance about much American sociological writing, the Institute pressed forward with ideas for new work. With the beginning of the World War II, the Horkheimer Circle formulated two projects that they thought might be of interest and use to the Institute's host country—a study of German history and culture that would uncover the roots of Nazism, and a study of anti-Semitism. Despite their efforts, which included elevating the profile of the Institute in the United States,[15] these initial bids for grant support failed. The grant proposals were too much like contributions to *Studies in Philosophy and Social Science*, the newly created English-language version of the Institute's journal. Although both were provocatively rich in unsubstantiated theoretical insights, they perplexed and puzzled their sociological readers. Both projects were formulated in too close proximity to Horkheimer's larger plans for a "Dialectics Project," which later became *The Dialectic of Enlightenment*.

While the Horkheimer Circle's approach was consistent with the general thrust of Critical Theory that had arisen from the traditions of Continental sociology and that had evolved during the first phase of American exile, it also anticipated the chilling disclosures of the *Dialectic of Enlightenment*. The first anti-Semitism proposal arrived at a hypothetical theory of prejudice that reinforced the Institute's pessimistic assessments of European humanism, enlightened Reason, totalitarianism, and mass culture. Instead of proposing an empirically verifiable image of the modern anti-Semite, the Horkheimer Circle's first published examination of the phenomenon was more concerned with locating this social threat within a broader societal critique of contemporary Western civilization. Despite the significance and importance of this work for the Institute and its theoretical endeavors, the anti-Semitism proposal of 1941 received little reaction from Americans. By not addressing or meeting the needs of American social scientists, the proposal represented an unusual approach to a topic of great importance. Until the Institute could adopt the "scientific" standards of American social researchers, the Horkheimer Circle would find little support for their plight or for Critical Theory in the United States.

Without the inspiration of Paul Lazarsfeld, it is highly doubtful that the successful grant from the American Jewish Committee (or AJC) would ever have been awarded to the Horkheimer Circle. Early during the Institute's collaboration with the Jewish Labor Committee and the American Jewish Committee, strategies for managing the project and cooperating with American sponsors were discussed, and Lazarsfeld was touted as the appropriate role model for the Horkheimer Circle. As Pollock explained in an internal memorandum, "I am told that that is the way how the Social Science Research Council, the Rockefeller Foundation and project directors, who know their job like Lazarsfeld, prepare the extension of projects."[16] While the Horkheimer Circle suffered a long string of disappointments in the United States, Lazarsfeld had a clear record of success. Lazarsfeld had arrived at the same time as the members of the Institute, but he had come to America with relatively little. In the same time that it had taken the Horkheimer Circle to jeopardize its existence, Lazarsfeld had rapidly climbed the academic ladder to a post at an American university and the directorship of the prestigious Princeton Radio Research Project. By the early 1940s when his friends at the Institute for Social Research desperately sought his help, the tables had entirely turned. The former work-for-hire research assistant had become a valued mentor and advisor to his former employers.

While Franz Neumann largely took it upon himself to lead the stubborn charge to obtain the necessary funding for the anti-Semitism project, the rethinking and revising of the project was a collaborative effort that involved the entire Horkheimer Circle, as well as its friends in America. Although the assistance from outside of the Institute came in varying degrees, a fascinatingly diverse group lent their support and opinions—such as Robert Lynd, Eugene Anderson, Robert MacIver, Paul Lazarsfeld, Daniel Bell, Benjamin Nelson, Moses Finkelstein (later known as M.I. Finley), Paul Tillich, Hadley Cantril, Harold Lasswell, and Gordon Allport. These allies in America read drafts of rewrites and shared their reactions. They helped to inform and remind Neumann and the other members of the Institute of the interests and assumptions held by their target audiences in the social sciences.

The revised proposal that was finally accepted by the American Jewish Committee is in stark contrast to the 1941 publication that appeared in *Studies in Philosophy and Social Science*. Although the revisions of the project did bear some resemblance to its initial articulation, the successful grant proposal highlights the extent of the Horkheimer Circle's willingness to intellectually assimilate. Instead of unabashedly maintaining the integrity of their early theoretical plans, the Institute reframed both itself and the project within an American context. Furthermore, this revised proposal bears the imprints of the many interlocutors that assisted the Institute for Social Research. Not enough of the correspondence exists for one to untangle the specific roles played by the various thinkers that shared their advice with the Institute, but the preserved documents do make it clear that the process was mediated by outside forces in a way that was both new and different for the Horkheimer Circle. Like all other cases of cultural and intellectual assimilation, this crucial shift in the history of the "Frankfurt School" was a social process played out between the members of the Institute and the "native" friends willing to offer assistance.

This effort at U.S. integration was evident from the outset of the new grant proposal. Unlike the earlier drafts of the anti-Semitism project that made little mention of American scholars or their work in the field, the revisions from the fall and winter of 1942 demonstrated the Institute's support from U.S. intellectuals, as well as its knowledge of recent American literature on the topic.Instead of introducing the proposal with a broad theoretical statement that expressed the group's interpretation of modern society, the successful AJC petition inserted a list of sponsors and testimonials to precede the introduction. The testimonials, in particular, performed the impressive function of highlighting both the importance of the project, as well as the Institute's capabilities. Representative is the following tribute from Harold Lasswell: "Few topics are more urgent than this and few institutions are of equal competence to your own for the successful prosecution of research. Your Institute is distinguished for careful study of individual and cultural processes and is admirably equipped to bring out the full complexity of the inter-relationships involved."[17] Especially when one considers the reputations of those being quoted, the impact on the readers from the AJC must have been significant. Here were some of America's prominent social scientists praising both the Horkheimer Circle and its research plans.

The actual introduction to the revised grant proposal further echoed the new image of the Institute that had already been sounded in the list of sponsors and testimonials. To underscore the prominence and importance of studying anti-Semitism, the Institute no longer restricted the discussion to its own theories about prejudice and the crisis of contemporary culture. Instead, the members focused on recent American writings on anti-Semitism and suggested that the rise in interest reflected a growing appreciation for the importance of the problem.[18] American commentators, according to the Horkheimer Circle, were increasingly becoming interested in the phenomenon, because anti-Semitism was undergoing a fundamental transformation from its premodern and archaic manifestations. Although few recognized what was happening, anti-Semitism was becoming totalitarian. Such a shift signaled new dangers to worldwide Jewry and justified the need for research into this frightening development.

The Horkheimer Circle still utilized a revised Marxian analysis to see the rise of totalitarianism as a contemporary crisis within capitalist societies. The primary threat, however, no longer jeopardized merely the working class, small businesses, and free professionals. The Horkheimer Circle broadened their rhetoric to present the totalitarian menace in terms that ordinary Americans would appreciate. It threatened liberty, democracy, and the middle class—the very foundations of American society.

As much as the new AJC proposal represented an assimilation to American interests and institutions, some elements of the project were envisioned to remain the same. Most notably, Germany was maintained as a primary focal point. Although the project reenvisioned the anti-Semitic menace to highlight its threat to the United States, National Socialism still served as the most advanced prototype of the totalitarian danger. Consequently, the Institute suggested:

> Since German anti-Semitism is more fully developed than any other, the case of Germany supplies us with the best point of departure, the best model for studying the social basis of anti-Semitism, its methods of propaganda, and the psychological mechanism with which it operates.[19]

Germany could function as an historical case study that might shed light on the circumstances that gave rise to totalitarian anti-Semitism, as well as possibly illuminating the potential consequences of its proliferation. Unlike the 1941 proposal, however, the Horkheimer Circle recognized the importance of moving beyond the lessons that could be taken from Germany. To meet this end, they adopted an infection model for totalitarian anti-Semitism that seemed to suggest the existence of potential parallels between Weimar Germany and the postwar conditions that were likely to exist in Europe and the United States. As the revised proposal suggested,

> After the collapse of National Socialism, the new form of anti-Semitism it has created is likely to survive it and spread to countries other than Germany [. . .]. After the war various social groups may try to regain their pre-war privileges or retain their war gains. Small and middle business men will fight for the restoration of their independence; the farmers will clamor for subsidies without which many of them would be reduced to starvation; big business will try to get rid of government interference—an attempt which will certainly be opposed, as it might create a wild post-war boom followed by a dangerous depression; labor will demand full employment, more social security, a share in the management and profits. Every group will try to shift the burden of the war costs on the other groups, and in this situation it is likely that a strong "social demand" for a vigorous anti-Semitic policy will arise.[20]

Like the 1941 proposal, the revised investigation promised to include a study of the origins and history of anti-Semitism. The description, however, was far less impressionistic. Although the earlier work proposed some intriguing but tenuous hypotheses, the new appeal to the AJC promised to rigorously uncover the history of prejudice and mass persecution. Instead of identifying specific events and intellectual movements that anticipated contemporary anti-Semitism and failing to appropriately explain the basis for each, the Horkheimer Circle outlined their procedure without sharing their social philosophical insights and expectations. The result was a coherent and concrete description of a research methodology that could be embraced by historians and social scientists in the United States. This historical analysis promised to accomplish what the prior grant proposal had offered (an historical examination of anti-Semitic behavior and its rhetorical basis in the messages of demagogues), but it devoted itself to these topics without drawing attention to the Continental orientation of Critical Theory and thereby undermining the credibility of the project.[21]

Historians of the "Frankfurt School" offer a variety of interpretations regarding this milestone in the Institute's U.S. history. All agree that *The Studies in Prejudice* represent a major shift in Critical Theory, but the explanations for this transition differ. Some, like Helmut Dubiel, minimize the importance of the change.[22] Dubiel prefers to de-emphasize the empirical research projects and instead focuses on the theoretical production of the *Dialectic of Enlightenment*. At the same time that Critical Theory had abandoned its revolutionary audience and the members of the Institute were fashioning philosophical and cultural critiques largely for themselves, there was mysteriously no attempt to mediate empirical research with theory, as was the case during the group's earlier history. Martin Jay, by contrast, convincingly suggests that the Horkheimer Circle grew increasingly more comfortable with empirical research as they learned more about it during their American sojourn.[23] Wiggershaus, meanwhile, proposes two possible explanations. Recognizing that anti-Semitism had grown into the focal point of the Institute's interests by the winter of 1942–1943, Wiggershaus believes that material and reputational needs forced the topic of anti-Semitism into the group's theoretical work (such as the dialectics project), or the work with the AJC served as an empirical complement to the group's philosophical speculations.[24] Coser, who worked with the members of the Institute on Morningside Heights, suggests that *The Studies in Prejudice* might be viewed as a reconciliation with the Institute's inherent bourgeois tendencies. All came from well-to-do German families, and they had always seen themselves as radicals in Babylonian exile. The anti-Semitism project therefore most notably suggested an abandonment of revolutionary utopianism and the temporary adoption of American liberalism.[25] Hohendahl and Tar, viewing the shift in more purely intellectual terms, emphasize the significant reorientation in sociological outlook. Whereas the Institute formerly focused on the socioeconomic underpinnings of social phenomena and the structural transformation of society, *The Studies in Prejudice* marked a move toward depth psychology and ideological reeducation.[26]

My investigation suggests that all of these interpretations are correct, and I only seek to amend them by perhaps recombining them to develop a more comprehensive picture of the Institute by looking at it more closely in relation to its cultural and intellectual surroundings. Clearly, the *Studies in Prejudice* do mark a major milestone that is partly the result of intellectual assimilation, de-radicalization, and professional necessity. The Institute's financial needs made a successful appeal to the AJC's a necessity for those in the Horkheimer Circle that did not want to see it broken apart (Horkheimer's initial plan for coping with the loss of the group's subsidies). At the same time, such a project was only imaginable after the orientation and interests of the group had changed. Thus, a topic such as anti-Semitism could become a main arena in which the group could pursue a whole myriad of interconnected themes. Many factors were at work in the coming of this major transition. Likewise, there were many people involved. Although Franz Neumann led much of the effort, he was closely assisted by Otto Kirchheimer, Herbert Marcuse, Leo Löwenthal, and Theodor Adorno. Similarly, there were all of the Institute's American advisors that played a major role in helping the Horkheimer Circle rethink the project in light of the interests and assumptions of the possible sponsors. Were there more documentary evidence, one could determine specific influences. Nevertheless, enough material does exist for us to recognize the powerful influence that Paul Lazarsfeld exerted—at first as a role-model and then later as a formal advisor to the *Studies in Prejudice*.

Despite his partly deserved reputation as a sociological mercenary and as a professional opportunist, Lazarsfeld developed a successful strategy for pursuing social research in America. Instead of creating theory for its own sake or only tackling those questions that interested him as a researcher, Lazarsfeld encouraged those around him to seek out socially relevant and methodologically acceptable topics that overlapped with one's theoretical, political and epistemological agendas. If applied successfully, this strategy could gain support from U.S. sociological foundations for topics that represented the true interests of the researcher. The revised anti-Semitism proposal that gained acceptance from the AJC in March 1943 was the Horkheimer Circle's application of Lazarsfeld's strategy. The successful AJC grant formulated an anti-Semitic threat that was relevant to American Jewry and envisioned a methodology that was consistent with U.S. sociological practices. In addition to rescuing the Institute's finances and reputation, the money that paid for the anti-Semitism project also provided Horkheimer and Adorno with salaries that enabled them to think more generally about the topic. Consequently, the Institute for Social Research concluded the 1940s with the successful publication of *The Studies in Prejudice*, as well as the completion of their theoretical masterpiece, the *Dialectic of Enlightenment*.

To underscore Horkheimer's own initial discomfort with the revised anti-Semitism project and the sociological methods that he would need to carry out in the wake of the project's funding by the AJC, it is important to consider some of his comments from the first months following the acceptance of the Institute's proposal to the American Jewish Committee. As Horkheimer admitted in a letter to Marcuse,

> The problem of Anti-Semitism is much more complicated than I thought in the beginning [. . .]. Since we have decided that here in Los Angeles the psychological part should be treated, I have studied the literature in this respect. I don't have to tell you that I don't believe in psychology as a means to solve a problem of such seriousness [. . .]. If we could succeed in describing the patterns, according to which domination operates even in the remotest domains of the mind, we would have done a worthwhile job. But to achieve this, one must study a great deal of silly psychological literature and if you could see my notes [. . .] you would probably think I have gone crazy myself.[27]

The letter to Marcuse indicates that Horkheimer and the members of his circle were not anticipating the acceptance of their proposal by the American Jewish Committee. More importantly it demonstrates their lack of preparation for the project on anti-Semitism and points to the extent to which they had relied on friends and advisors like Lazarsfeld in the drafting of the successful grant proposal.

The tragic irony of this clear case of émigré cooperation and intellectual assimilation was that the Horkheimer Circle's success came too late, and it was their collaborator who displaced them. By the time Lazarsfeld and others were assisting the Institute in its final appeal to the American Jewish Committee, Columbia's sociologists had recognized Lazarsfeld and his radio group as the realization of their persistent quest to bring an empirical research Institute to Morningside Heights.[28] Like the Institute for Social Research, Lazarsfeld's group was financially self-sufficient. What set Lazarsfeld apart and made him preferable was that he had a proven track record in America. There was little question that Lazarsfeld would alter Columbia and build it into a center for quantitative social research. Thus Lazarsfeld was brought to Morningside Heights to fulfill Robert Lynd's desire for an empirical researcher, while Robert Merton was hired to satisfy Robert MacIver's desires for another social theorist in the department. This compromise not only healed the rift among the sociology department's faculty, but also revived the former glory of the institution by creating the kind of reinvigoration that the department had hoped for in the late 1920s.

Following his appointment at Columbia, Lazarsfeld still tried to remain a friend of the Institute for Social Research. He continued to employ many members of the Horkheimer Circle, he provided assistance and advice on nearly every aspect of the *Studies in Prejudice,* and he defended the Institute against its detractors on Morningside Heights.[29] It would be a mistake, however, to see this support as entirely friendly or altruistic. Lazarsfeld had not risen so quickly without being a very savvy administrator. He promoted the Horkheimer Circle, but not always with its best interests in mind. By the end of the 1940s, Lazarsfeld put together a deal that would have brought nearly all of the members of the Horkheimer Circle to Columbia. In retrospect, however, it is impossible to see the scheme as anything but naked opportunism. Knowing of the wedge that Horkheimer had driven between the California and New York branches of the Institute, Lazarsfeld suggested that the Bureau for Applied Social Research could absorb all but the most theoretically oriented members of the Institute—thus excluding the inner circle of Horkheimer, Pollock, and Adorno. A huge portion of the Institute's endowment would be granted to Lazarsfeld, and the proceeds would be used to pay for the salaries and research of the Institute's New York branch. Not surprisingly, Horkheimer swiftly rejected the offer and opted instead to return his faction of the Institute to Frankfurt.[30]

Both parties prospered in the long run, but it is not hard to understand the strained and complicated relations that existed between the two groups. The negative energy that arose from Lazarsfeld's proposed partition of the Institute set this dynamic in motion, but the tensions persisted throughout the postwar period. These feelings were only magnified as former castaways from the Institute, such as Franz Neumann and Leo Löwenthal, remained closely connected to Columbia and the Bureau for Applied Social Research. Although the two institutions continued to cooperate, the awkwardly entwined legacies of their central figures drove a sharper wedge between them. For the remaining members of the Horkheimer Circle, Lazarsfeld was a reminder of their mistakes and failures during exile. Perhaps, more importantly, he also highlighted their unsteady assimilation of American research methods. Despite their constant worries about the spreading influence of Positivism in the social sciences, the members of the Horkheimer Circle did return to the Federal Republic as ambassadors of both American sociology and its empirical methods. Furthermore, their postwar success relied on their complex status as political and cultural "outsiders" recast in the wake of the war as "insiders." Their diagnosis of the totally administered society required some degree of notoriety in the same civilization that was the basis of this critique. Thus, the Horkheimer Circle had to succeed as "operators" for their theories regarding the "operator society" to gain as wide an audience as they did.

Lazarsfeld suffered an understandable discomfort when scholars and historians in his new home began focusing on the "Intellectual Migration" at a time when Critical Theory was at the height of its popularity. In the wake of Marcuse's international celebrity, the follies of the "Frankfurt School" were forgotten amid the newfound fascination with Hegel, the Young Marx, and Western Marxism. For Lazarsfeld, who had attained a degree of success in his field that was unusual within his émigré cohort, this surprising twist came with its own betrayal. As Critical Theory gained more attention, its critique of Positivism was recognized for what it partly was—a swipe at the U.S. social scientists that had helped the Horkheimer Circle during its most desperate days in exile. As Lazarsfeld wrote,

> When, after the war, the majority of the Frankfurt group returned to Germany, they at first tried to convey to their German colleagues the merits of empirical social research which they observed in the United States [. . .]. Within a period of five years, however, the situation changed completely. Adorno embarked on an endless series of articles dealing with the theme of theory and empirical research. These became more and more shrill, and the invectives multiplied. Stupid, blind, insensitive, sterile became homeric attributes whenever the empiricist was mentioned [. . .]. Thereafter one paper followed another, each reiterating the new theme. All have two characteristics in common. First, the empiricist is a generalized other—no examples of concrete studies are given. Second, the futility of empirical research is not demonstrated by its products, but derived from the conviction that specific studies cannot make a contribution to the great aim of social theory to grasp society in its totality. Empirical research had become another fetish concealing the true nature of the contemporary social system.[31]

In the wake of Critical Theory's unanticipated success, Lazarsfeld was more content to sever his legacy from the "Frankfurt School" than to set the historical record straight. In retirement, he was willing to promote the mythic image of the "Frankfurt School" that had emerged with its reception by the New Left rather than to permit his work and reputation to be linked to this intellectual phenomenon. Thus, we are left with these contested reputations and legacies, which bear little resemblance to the archival record.

Notes

1. Helmut Dubiel, Theory and Politics, *Studies in the Development of Critical Theory* (Cambridge, MA: The MIT Press, 1985); Martin Jay, *The Dialectical Imagination: A History of the Frankfurt School and the Institute of Social Research, 1923–1950* (Berkeley, CA: University of California Press, 1996); David E. Morrison, "Kultur and Culture: The Case of Theodor W. Adorno and Paul Lazarsfeld," *Social Research* 45, 2 (1978): 331–355; Rolf Wiggershaus, *Die Frankfurter Schule* (Munich: Carl Hanser, 1986).

2. Theodor W. Adorno, "Scientific Experiences of a European Scholar in America," in Donald Fleming and Bernard Bailyn (eds.), *The Intellectual Migration: Europe and America, 1930–1960* (Cambridge, MA: Harvard University Press, 1969): 338–370; Paul Lazarsfeld,

"An Episode in the History of Social Research: A Memoir," in Donald Fleming and Bernard Bailyn (eds.), *The Intellectual Migration: Europe and America, 1930–1960* (Cambridge, MA: Harvard University Press, 1969): 270–337. By bracketing the term "Frankfurt School," the author seeks reserve questions about its problematic historical authenticity. I substitute the term "Horkheimer Circle" in my effort to move beyond conceptions of the Institute for Social Research that arose long after their emigration from Germany and return to Frankfurt.

3. Theodor W. Adorno et al., *The Positivist Dispute in German Sociology*, trans. Glyn Adey and David Fridby (New York: Harper and Row, 1976).

4. See Lazarsfeld's memorandum entitled "Foreign Research Service of the Bureau of Applied Social Research," dated 4/27/50 from the *Paul Lazarsfeld Papers*, Rare Book and Manuscript Library, Butler Library, Columbia University, box 6, file 18.

5. Thomas Wheatland, "Isolation, Assimilation, and Opposition: A Reception History of the Horkheimer Circle in the United States, 1934–1979" (Ph.D. diss., Boston College, 2002).

6. Lewis Feuer, "The Frankfurt Marxists and the Columbia Liberals," *Survey* 25, 3 (1980): 156–176; Martin Jay, "Misrepresentations of the Frankfurt School," *Survey* 26, 2 (1982): 131–141; G. L. Ulmen, "Heresy? Yes! Conspiracy? No!" *Survey* 26, 2 (1982): 142–149; Lewis Feuer, "The Social Role of the Frankfurt Marxists," *Survey* 26, 2 (1980): 150–170.

7. See "Memorandum on the Social Science Needs of Columbia University," written and circulated during 1928–1929, *Samuel McCune Lindsay Papers*, Columbia University Rare Book and Manuscript Library, box 46.

8. "The Constitution of the International Society for Social Research," prepared by the law firm of Mitchell, Taylor, Capron and Marsh, signed and ratified by Max Horkheimer and Friedrich Pollock in Geneva on 2/25/34. From *The Columbia University Archives* (formerly known as the Columbiana Collection), file folder 549/7.

9. Max Horkheimer, "The Present Situation of Social Philosophy and the Tasks of an Institute for Social Research," *Between Philosophy and Social Science: Selected Early Writings*, trans. G. Frederick Hunter, Matthew S. Kramer, and John Torpey (Cambridge, MA: Harvard University Press, 1995), 1–14.

10. See "Notes for a Talk," dated 1934 from the *Max Horkheimer Archiv*, Stadt und Universitätsbibliothek, Frankfurt-am-Main, box series IX, folder 52, document b. Gumperz's name is written at the top of the first page in Horkheimer's handwriting.

11. See Robert Lynd "Memorandum on the Study of Changing Family Patterns in the Depression," dated 3/14/33 from the *Robert Lynd Papers*, Library of Congress, microfilm reel 2.

12. See Robert Lynd "Possibilities of Cooperation among Research Institutions," presented as a talk at a meeting of the SSRC held at the Brookings Institution, in February 1931. From the *Robert Lynd Papers*, obtained on microfilm from their archival repository at the Library of Congress, Manuscript Division, microfilm reel 2.

13. See the *Samuel McCune Lindsay Papers*, boxes 59–64, which indicate the widespread involvement of Columbia sociologists in the Newark study, and see Mirra Komarovsky, *The Unemployed Man and his Family: The Effects of Unemployment upon the Status of the Man in Fifty-Nine Families* (New York: Institute for Social Research, 1940).

14. See the letter from Beatrice Doerschuk, Sarah Lawrence's Director of Education, to Erich Fromm dated 3/9/39 from the *Erich Fromm Papers*, obtained on microfilm from the collection of Fromm's papers formerly kept by the New York Public Library, microfilm reel 1, and see Ruth Monroe, *Teaching the Individual* (New York: Columbia University Press, 1942).

15. Partly this was accomplished through professional networking, but (in an ironic twist of fate) the famed critics of the "culture industry" also used the services of a P. R. firm—the Pheonix News Publicity Bureau. See Horkheimer's letter to Nicholas Murray Butler dated 12/1/38 from the *Max Horkheimer. Gesammelte Schriften. Band 16*, ed. Gunzelin Schmid Noerr (Frankfurt: Fischer, 1995), letter number 443, 516–517.

16. See Pollock's "Memorandum re: Antisemitism Project" dated 10/29/43 from the *Max Horkheimer Archiv*, box series VI, file 34, 146–147.

17. "Excerpts from Testimonials," from Institute for Social Research, "A Research Project on Anti-Semitism," from the *Max Horkheimer Archiv*, box IX, file 92, document 7a.

18. The Horkheimer Circle cited articles published in 1941, such as D. W. Petregorsky, "Anti-Semitism: The Strategy of Hatred," *The Antioch Review* 1 (Fall 1941); Michael Straight, "The Anti-Semitic Plot," *The New Republic* 105, 2 (Sept. 22, 1941); Benjamin Akzin, "The Jewish Question after the War," *Harpers Magazine* 1096 (Sept. 1941); Albert J. Nock, "The Jewish Problem in America," *The Atlantic* 167, 6 (June/July 1941); James Marshall, "The Anti-Semitic Problem in America," *The Atlantic* 168, 2 (Aug. 1941); Arthur H. Compton, "The Jews: A Problem or an Asset," *The Atlantic* 168, 4 (Oct. 1941); Miriam Syrkin, "How to Solve the Jewish Problem," *Common Ground* 2, 1 (Fall 1941). See "An Introductory Statement," Ibid, 1–2.

19. Ibid., 3.

20. Ibid., 17–24.

21. Ibid., 8–9.

22. Helmut Dubiel, *Theory and Politics: Studies in the Development of Critical Theory* (Cambridge, MA: The MIT Press, 1985): 106.

23. Martin Jay, *The Dialectical Imagination: A History of the Frankfurt School and the Institute of Social Research, 1923–1950* (Berkeley, CA: University of California Press, 1996): 221–224.

24. Rolf Wiggershaus, *Die Frankfurter Schule* (Munich: Carl Hanser, 1986), 320–321.

25. Lewis Coser, Refugee Scholars in America: Their Impact and Their Experiences (New Haven: Yale University Press, 1984): 97.

26. Peter Uwe Hohendahl, *Prismatic Thought: Theodor W. Adorno* (Lincoln: University of Nebraska Press, 1995), 41–52; Zoltan Tar, *The Frankfurt School: The Critical Theories of Max Horkheimer and Theodor W. Adorno* (New York: John Wiley and Sons, 1977), 102–112.

27. See the letter from Horkheimer to Marcus dated 7/17/43 from the *Max Horkheimer Archiv*, box series VI, file 27a, documents 12–13.

28. See the reports entitled "About Columbia University's Bureau of Applied Social Research" and "Columbia University's Bureau of Applied Social Research: Its Objectives and Purposes," from the *Paul Lazarsfeld Papers*, Columbia University Rare Book and Manuscript Library, box 6, file 18. Also see the reports of Columbia's Cheatham Committee that examined the rise, role, and impact of the Bureau of Applied Social Research from the *Paul Lazarsfeld Papers*, box 6, file 16.

29. See Lazarsfeld's notes and recommendations for the Horkheimer Circle's unpublished "Anti-Semitism Among American Labor," in the *Paul Lazarsfeld Papers*, box 20, file 1, as well as box 36. Also see his comments on *The Authoritarian Personality*, box 20, file 13. Although Lazarsfeld's defense of the Institute before Columbia's Sociology department contained a few back-handed compliments, he was strongly supportive of the Horkheimer Circle and its work—see Lazarsfeld's letter to Theodore Abel dated 2/5/46 from the *Max Horkheimer Archiv*, box series II, file 5, document 149.

30. See the letter from Lazarsfeld to Horkheimer dated 5/21/46 and the response from Horkheimer to Lazarsfeld dated 6/10/46 from the *Max Horkheimer Archiv*, box series II, file 5, document 149.

31. See Paul Lazarsfeld's article entitled "Critical Theory and Dialectics," from the *Paul Lazarsfeld Papers*, box 36, 112–114. Originally, this paper was delivered as a talk and was first published in "Main Trends of Research in the Social and Human Science," (Paris: Mouton, Unesco, 1970): 111–117. It was also published in Paul F. Lazarsfeld *Qualitative Analysis: Historical and Critical Essays* (Boston: Allyn and Bacon, Inc, 1972): 168–180.

Chapter Thirteen

"Political Culturalism?" Adorno's "Entrance" in the Cultural Concert of West-German Postwar History

Alfons Söllner

Adorno's increased presence in the media on the occasion of his hundredth birthday, the amplification of his twenty-volume collected works by a seemingly endless stream of posthumous papers, the popularization of an esoteric thinker through biographical studies made possible above all by the availability of his extensive correspondence[1]—all this cannot disguise the fact that there remains a puzzle about "Adorno in the Federal Republic of Germany." Politics and society were reviving, and yet they were also as if lamed by the leaden weight of the past. Only in the sphere of culture did numbed looks backward intersect with hopeful looks forward to create an opening for understanding. How was it possible for a Jewish re-emigrant to carry his existential confrontation with the society that had driven him away into the daily routine of the professional role he now assumed? In view of this starting point, it is astonishing to consider the sheer inexhaustible productivity of an author, who contributed to nearly every category encompassed by bourgeois high culture, feuilleton publication and scholarly work, and who became, in the course of a mere two decades, the archetypical Leftist intellectual of German postwar history. Even more surprising is the growing and eventually extraordinary success that this man had in a society for which he, after all that had happened, could have had as little liking as it had for him—a Jewish re-emigrant, who had yet to find his place in the uncertain and divided ground of West German postwar culture.

How was this development possible? What can it tell us about an intellectual in an exposed position, who always—even when he acts as its harshest critic—is a mirror of his field of operations? And what can it tell us, on the other hand, about the situation of society, about its political tendencies and cultural dynamics, which need not by any means run in parallel? Through questions of this sort, and above all through an attempt at their methodological linkage, the following study departs from the conventions of intellectual biography and turns instead to the ambiguous field of political culture. If political culture as a field of research is inherently uncertain, it is all the more so in the case of the young Federal Republic. If anything at all can be said to define this epoch, then it is its transitional character, a complex interplay of continuities and new beginnings, one could even say a simultaneity of diametrically opposed tendencies, of which the looming presence of the totalitarian past and the orientation toward a democratic future form the outermost poles. This chapter aims to describe and more thoroughly plumb the depths of this disorienting jumble of culture and politics. One way to this end leads perhaps through an understanding of how an intellectual re-emigrant like Adorno found his bearings in this specific configuration, and what he was able to make of it.

The thesis to be explored in this work is methodologically as well as materially somewhat paradoxical: it assumes that the initial political–cultural position of the Federal Republic of Germany consisted of a certain amalgam of contradictory elements, the spectrum of which stretched from inherited cultural conservative inclinations to decisive intentions toward political modernization. An intellectual who wished to be successful in this milieu must not only have precisely understood this difficult configuration, but he must also have possessed a key with which to unlock it. Adorno established himself in the early history of the Federal Republic above all as a brilliant and universal cultural critic, and through this he developed a virtuosity, from which perhaps the mystery of the origins of his success in the political sense can be deciphered. Even as he established himself definitively as an advocate of the modern in music, literature, and academics, he remained at the same time committed to a normative understanding of culture that was highly selective, esoteric, and elitist. Adorno was, to a much greater degree than the cliché of the Leftist intellectual implies, bound to the tradition of the German religious conception of art. Here the most interesting question is whether it was in fact precisely with this explosive mixture that he not only conquered the hearts of his audience, but also stimulated it to political responses—indicative of enormous potential—that were able to further the development of the youthful Federal Republic.

In seeking to answer these questions, in this chapter I concentrate exclusively on the beginning of Adorno's career in the Federal Republic. While this initial period constitutes only a brief and ephemeral moment around 1950, this initial situation has much significance for later developments, since it already sounds all the decisive motives, which Adorno only elaborates and, above all, orchestrates compellingly in composing the extensive body of his subsequent work.

Adorno returned to Germany as a fully formed intellectual and quickly found his part in the developing cultural concert of postwar history. What was a mere confusion of voices for others, Adorno understood as his "entrance," which I analyze first through two exemplary texts from the year after his return: an assessment of the situation and a programmatic statement. In the second part, I show how Adorno begins to transform the central leitmotivs of the history of his life and thought, and thereby develop particular methods of argumentation and writing, which I call "political culturalism." The chapter in its entirety attempts to elucidate this concept step by step, in order to make the proposition plausible, in conclusion, that a key to Adorno's exceptional influence as a political intellectual could lie here, in using cultural means for political ends.

Assessment of the Situation and Cultural-Critical Program

When Adorno traveled to the recently established Federal Republic in the autumn of 1949, it was not at all certain that this would indeed be a return to Germany. He had only come to the University of Frankfurt as temporary replacement for Max Horkheimer, who had been called back to his former professorial position, which was itself quite a singular phenomenon in the reconfiguration of the West German universities after the war. In addition, both Adorno and Horkheimer attempted to maintain their rather loose American ties for as long as possible, since in Germany neither man had claim to connections that were firm or dependable enough to effect a final decision to remain. The universities, after all, showed that they knew how to cut themselves off to a great degree from denazification. Horkheimer continued to consider returning to America, in fact, even after the establishment of the Institute for Social Research; and in the case of Adorno, the decision to stay seems to have first become final after an extremely negative experience during an extended research trip to Los Angeles in 1951/52—it was to be his last stay in the United States.

When one examines the documents from the time around 1950, one is struck by a certain ambivalence, and it is this that provides a point of departure for our specific interests. Although critical observations and skeptical opinions about Germany dominate his private letters to Horkheimer and to the "highly esteemed" Thomas Mann, the published essays from this same time are much more constructive and optimistic. We will most clearly grasp Adorno's initial position if we do not limit ourselves to the monographs with which Adorno entered into the West German postwar period. These manuscripts had already been more or less completed in emigration and thus reflect that position in a way that will later be of concern to us. More significant at this point are the texts directly connected to the return itself, those that were written in Germany—so to speak, from within. Here I have selected two texts, one of which recounts the impressions of a re-emigrant who must get his bearings in an entirely new situation, while the other encapsulates and brings up to date the theoretical premises that Adorno had brought with him from his time in emigration.

Adorno took up his professorial position in the autumn semester and began setting down his first impressions in letters to America. The details can now be found in the edited correspondences with Thomas Mann, Max Horkheimer, and Adorno's parents.[2] During the winter break, however, his notes seem to have grown into a report on his experience, which he then published in the *Frankfurter Hefte* in May of 1950 under the title *Resurrected Culture*[3]. What is the tenor of this text, and what does it reveal about the perspective of the engaged returnee? It begins in a way contrary to what one would expect. It had been a topos of emigration that a genuine cultural life could not exist under Hitler, which certainly didn't bode well for the period thereafter. One finds this sort of thing already in *Minima Moralia*, which was written between 1944 and 1947. In *Resurrected Culture*, however, Adorno makes the following observation:

> After long years spent in emigration, the intellectual who returns to Germany is surprised straight away by the intellectual climate [. . .]. One reckons on the breakdown of culture, on the disappearance of participation in anything that goes beyond daily concerns [. . .]. But that is simply not the case. The relationship to intellectual things, understood in the very broadest sense, is intense.[4]

The "students' passionate engagement with relevant questions" pleased Adorno and transplanted him to the early Romantic period, "during which one counted a book as little popular as Fichte's *Wissenschaftslehre* among the greatest events of the age."[5] Adorno extended the eager engagement that had so impressed him among his students at the University of Frankfurt to Germans in general, and thus believed himself able to make a virtue of their necessity in the aftermath of National Socialism: he sympathetically saw their cultural "ravenousness" as derivative of the "intellectual barrenness of the Third Reich."[6] The post-war German cultural situation appeared not only as developmentally delayed in comparison to the culture industry he had experienced and suffered under in America, but also as isolated. This isolation, however, carried concealed within it a special opportunity: "It has not yet gotten around that culture in the traditional sense is dead [. . .] Contact with culture in post-war Germany has something about it of the dangerous and ambiguous comfort of hiding away in the provinces." And so Adorno not only feels himself made at home and yet somehow restricted by it, but he also saw the larger perspective: "The more mercilessly the world-spirit triumphs, the sooner will that which remains be able not only to stand in for the lost, for the romantically obscured past, but also to reveal itself as an escape route and a place of refuge for something better in the future."[7]

Of interest to us now is how Adorno intended to emerge from this thoroughly precarious situation. For the moment, he could only grapple with that which posed an obstacle to the obvious desideratum of radical intellectual rebirth. Included in this, for example, is his observation that even in young people "intellectual energy is transferred in a quite curious way onto questions of interpretation of existing visual art, poetry or philosophy."[8] The complete ambivalence of this finding is further revealed by a historical comparison with the post-1918 period in Germany. Although at that time expressionism came on the scene as a brilliant and momentous artistic coup, today nothing even comparable to it can be found in Europe: "Even in the birth-place of aesthetic modernity, in France, there is no *avant-garde*." What gets carried over from there into a defeated Germany is much more of a modish existentialism, which blends together with the cultural remnants of the dismal interwar period into a heroic neo-humanism. The depressing result is the strange condition of an, as it were, "neutralized culture": "When compared to that which it [i.e., the self-conscious I of the expressionistic generation] is supposed to express, the art that fills the vacuum today appears as imitative or helpless or both. It is a ghostly traditionalism without a binding tradition."[9]

What is to be done about this? What is the task at hand? Adorno speaks of the "necessity of intellectual reorientation," yet for the initial position of interest to us, nothing is more informative than the fact that Adorno had little more than an abstract and hesitant answer to this decisive question. It is as if he wished to extend in a certain direction the knowledge of cultural theory with which he had returned, but then he immediately retracts his intentions. The sphere of action of the re-immigrant in postwar Germany is unavoidably politically conditioned, subjectively as well as objectively, but Adorno wished to remain in the familiar field of culture, even if it was not obvious, without reflection, that the possibilities there could be extended into the field of politics. In this spirit, he adopts Max Frisch's phrase of "culture as alibi": "*Bildung* today has not least the function of making forgotten and repressing the horror of that which happened and one's own responsibility for it."[10] Yet these generally clear formulations provide no political handhold, since Adorno remains somewhat unclear and general where he wishes to be concrete. And so he urges his reader to an "insight into the laws that brought about the disaster of the recent past," and trots out such vague tired phrases as "the concept of world's humane orientation and its theoretical establishment," and advocates in hortatory terms "the analysis of the actual current potential for the complete, substantial realization of freedom." The negative is much more graspable: "It is as if the human race were held under a spiritual spell. The lack of freedom and the belief in authority, even if it were only in the authority of that which now is, have made their way into general consciousness. No one quite trusts himself to address the distress and burning, and yet all are in truth aware of it."[11]

And as if to make up for his lack of concreteness in this description of the situation, which is certainly sensitive, but is also culturally limited, Adorno does in the end come to speak of the political situation in a narrower sense, if only in order to locate a reason for the absence of cultural modernity, for the "lack of explosive force, of passion for adventure, even of curiosity."[12] He goes so far in fact as to produce a direct connection between global political developments and cultural stagnation, and not only for the defeated and moreover divided Germany, but also for Europe in its entirety. Here Adorno has assembled a few keen insights that politically synthesize that "which for one returning from America melds into a unity in such a puzzling way." In this vein, he remarks: "From a political and anthropological point of view, it appears to me that what determines the present condition of the spirit here is the premonition that Germany has ceased to be a political subject in the long-familiar sense of a nation-state [. . .]. The paralysis of spiritual productivity is caused by the fact that one is collectively no longer a political subject and that one therefore does not undertake anything unbounded in the realm of spiritual reflection either. One accommodates oneself to the constellation of great powers and thinks that one can salvage some uniqueness only by resigning oneself to the delimited sphere of culture."[13]

If this diagnosis amounts to an insightful stretto composed of politics and culture, and if it serves, through this converging of the two—but indeed also through its revealing biographical location—as an early key documentation of Adorno's hidden potential, what conclusions did he draw from it? Do they allow one to read the traces of an argument that continue in other writings, that reveal a greater line, by which one could reconstruct Adorno's path from the hour of his return in the Federal Republic of the 1950s? As cautious as one must be in applying programmatic perspectives to a figure like Adorno—a thoroughly self-willed and at the same time intrinsically motivated nonconformist—the key words are clearly and urgently "culture" and "cultural criticism." This of course applies not only to Adorno's work after his return to Germany, but in fact characterizes the entire developmental history of this gifted offspring of the educated classes of Frankfurt and Main. But cultural criticism can now be taken as the dynamic center of operations, from which one can understand his extensive intellectual activity far into the 1950s and 1960s. We are thus justified in reaching for another text, which—no less problematic or idiosyncratic—is representative of Adorno's initial situation in postwar Germany.

Here I refer to the essay *Cultural Criticism and Society*, which Adorno wrote in 1949 and published in 1951 in a *Festschrift* for the sociologist Leopold von Wiese. His 1955 use of a lead article bearing the same subtitle in his first collection of essays, *Prisms*, shows the continued and pervasive importance of this topic.

This essay is significant also for an additional reason. In its conclusion we find the sentence, which has become the most famous of Adorno's standard quotations, so famous in fact, that its author felt forced to take it up again later and eventually, in his central philosophical work, *Negative Dialectic*, to retract it,[14] in order to protect the work and himself from misunderstanding: "Cultural criticism finds itself faced with the final stage of the dialectic of culture and barbarism. To write poetry after Auschwitz is barbaric. And this corrodes even the knowledge of why it has become impossible to write poetry today."[15] It is not the long reverberation of this phrase in the history of the Federal Republic that is of interest to us here, but rather a logical extrapolation, which at the same time constitutes a search for the explosive point that allowed for a cultural verdict to become a political signal. If it is possible to demonstrate through this example the interplay of esoteric formulation and exoteric effect and especially the mutual dynamics of both elements, then we will not only have learned something about Adorno's particular way of thinking and writing, but we will also have gained the confidence necessary to a better understand the tricky relationship of culture and politics in the young Federal Republic—its "political culture."

When reading the text from the beginning, however, a feeling of disillusionment quickly sets in, for after long-winded reflections on cultural criticism and society, the incriminating passage seems not merely displaced, but downright erratic. In an enigmatic formulation, the "questions of lyric poetry" are made responsible for the articulation of a powerful position of engagement, the political sense of which one can only guess. On the whole, the argumentation of the essay is as circuitous as it is pretentious. Rather than delineating and decoding culture as a subcategory of sociology, which involves a specific field and specific methods, Adorno begins with a polemic. He attacks the conventional self-understanding of the cultural critic. This critic, Adorno determines, has a usurper's relationship to his object, he claims the privilege of cultural possession, is victim of an overextension of his claims, and becomes addicted to a "belief in culture as such," which—and here is the about-turn of the argument—in a certain sense places him on the same level with his declared opponent, the fascist disdainer of culture. In making this move, the traditional cultural critic misses in any case not only his object, but also his professional purpose; because: "Criticism is an indispensable element in a self-contradictory culture, just as truthful in its lack of truth as the culture is untrue."[16]

Paradoxical plays on words such as these are quite common in Adorno's text, which is not frugal in its expenditure of forceful overgeneralizations and oversubtle theses. Although from a present point of view this may well be considered the expression of a bewildering dialectic, and one that leads to lumbering sentences and to assertions one could hardly confirm, we must not discard Adorno's concerns as a whole as incomprehensible. One must remind oneself of the presuppositions, to which he can only allude in the text, and which are of considerable consequence. Adorno not only places himself explicitly in the tradition of Marxist thought, which understands culture first and foremost as a formation of ideology, but he also generalizes this connection and at the same time draws quite radical conclusions from it. Culture—for him this is no longer merely the sum total of false consciousness, in the sense of Marx and Georg Lukács, and no longer the malleable "putty" of high and late bourgeois society, which the early Horkheimer circle attacked with ideological-critical and above all social–psychological methods. For Adorno, culture had become more self-enclosed—and here he reaches back to considerations he had been developing since the 1940s in conjunction with Horkheimer—"But the greatest fetish of cultural criticism is the notion of culture as such"[17]—as well as more ossified. It turns into its opposite and "gravitates toward mythology."[18] This assertion is followed, and without circumlocution, by the statement: "The cultural critic nourishes himself on the mythical obduracy of culture."[19]

The course of Adorno's reasoning is surely more complex than this brief reconstruction can show, but what is most striking about his attempt to comprehend the interrelation of culture and society is his radical inversion of this no less radical concept of self-enclosure. The acceptance of a direct analogy between societal and intellectual development becomes possible through recourse to a second fundamental premise. Adorno returns abruptly to the philosophy-of-history construction of *Dialectic of the Enlightenment,* upon whose axis of thought the relationship between Enlightenment and myth had been turned upside down. What is the decisive historical tendency? While in the bourgeois age, an element of critique, of utopia, vis-à-vis society was discernible in culture, today this discrepancy, where ideology-critique could find a foothold, has shrunk away. And this opens the way to an allusion to the "culture industry": it is the most modern and most developed form of ideology, of the completely dedifferentiated "entanglement of culture in commerce."[20] In a certain way, culture was always ideology, that is, a societal function, but now it is totally integrated. This changes the sense as well as the method of cultural criticism. Because there is no longer any apparent distinction between culture and society, the following conclusion is unavoidable: "The procedure of cultural criticism itself is the object of permanent criticism."[21] And that, in turn, indicates that the hour has struck for the advent of the theory that Adorno himself advocates: "What distinguishes dialectical from cultural criticism is that the former intensifies the latter to the point of superceding (*Aufhebung*) the very concept of culture."[22]

Adorno's reflections, it should be noted, are much richer in presuppositions and more complex than can be shown here. He goes back to the methodological alternatives of immanent and transcendent criticism—as Karl Manheim had very ambitiously formulated them in the mid-1920s—but only to show that they were now out of date, that they could no longer be considered adequate for the current state of historical development. Most important here is the conclusion, which is at first methodological: "Cultural criticism must become social physiognomy. The more the whole—divested of all natural elements, socially mediated and filtered—is 'consciousness,' the more it becomes 'culture'[. . .]. Today, ideology means society as appearance."[23] There follows a further pirouette of thought, in which judgment of reversal is extended to the relationship between society and culture itself—with a logic that is, as one can imagine, quite overgeneralized: "Ideology, the socially-necessary appearance, is today the society itself, insofar as its integral force and inevitability, its overwhelming existence-in-itself, points to the meaning which that existence has exterminated."[24] And so the circle drawn from the bird's eye view of philosophy-of-history speculation closes, and leads back in this way to an immanent criticism of culture.

The confusion seems complete, but there is method to it. If one wants to learn something about the condition of society, if one wants to decode its most developed state, one must not practice ideological criticism, but immanent criticism. "Immanent criticism of intellectual and artistic phenomena seeks to grasp, though the analysis of their form and meaning, the contradiction between their objective idea and that pretension, and to identify what is expressed in the consistency or inconsistency of the formations as such about the constitution of the existent."[25] This is not merely a supercession (*Aufhebung*) of the division of labor between cultural criticism and societal analysis; cultural criticism—in it immanent form, to be precise—appears as the sole remaining legitimate form of societal analysis. The question then becomes, which "spiritual formations" can make the extrapolation of conclusion from cultural criticism to society truly compelling. Obviously it cannot be a matter of indifference what the cultural critic who seeks to keep up with the times makes the object of his reading, speaking, and writing. Are there, then, privileged cultural products and spheres of culture that enjoy an open or secret alliance with the travails of the search for truth—whatever this may be?

It is interesting to note that Adorno seems for now to leave this decisive question open at the end of his programmatic essay, for example when he writes: "In the open-air prison which the world is becoming, it is no longer so important to know what depends on what, since everything is so much one and the same. All phenomena are fixed in place, like emblems of the absolute rule by that which is."[26] But then he changes his mind—and this is exactly the tender spot where the entire argument is suddenly blown up and the falling wreckage compacts into a traumatic experience, which is however made present only by means of metaphor—a culturalist wrapping of a political message. This culturalist wrapping bursts forth from the circular reflection on method and it is presented, one would like to say, as "existential," if the term did not come from an intellectual context that Adorno strongly rejected, Heidegger's *Jargon of Autheticity*. What in any case becomes tangible here is a first and explosive example of what I propose to analyze in this chapter as Adorno's "political culturalism." Tangible here in rudimentary form is the typical double strategy indicated by the following concept: culture has become entirely ideology, a part of the bad society. This "spell" can only be exploded by pronouncing the password. Once it has been pronounced, however, the shock is so great that the element that triggered it must be repackaged, veiled in culture.

Auschwitz as a Political Cryptogram—Adorno's "Cultural Politics" in the 1950s

Is there an aspect of disguise contained in the literary form itself in Adorno's distinctive entrance into the cultural scene of West German postwar history? Was this cultural-istic wrapping, which unquestionably characterized his entire undertaking, merely a biographical idiosyncrasy, or did it put into effect, whether or not it was so intended, a hidden political strategy that availed itself of these cultural means and materials?

This brings us again to the question of "political culturalism." It can certainly be affirmed if we approach our theme from the perspective of a history of influence. It is certain that this culturalist wrapping, and in particular the element of defensiveness that it contained, are among the factors that won Adorno a largely positive reception.[27] The measured doses of a "working through the past" as which one could also ingest the moral–philosophical miniatures of *Minima Moralia* were in no way the sternly prescribed political Enlightenment represented by the early program of denazification and in the later practice of reeducation. If the problems of guilt and morality associated with National Socialism could be garbed in the guise of such "questions of conscience" as whether one should, for instance, call Stravinsky a fascist, when he too had been forced to emigrate, then the bourgeois audience could also face up to unpleasant questions about the immediate past. After all, it was not a matter of such concrete things as professional involvement in the Hitler Regime or questions of criminal responsibility.

On these premises, how does what was earlier analyzed as the desideratum of an "intellectual reorientation" in postwar Germany appear? The reply is that it remains no less negative. A significant note in *Minima Moralia* reads as follows: "The idea that life after this war will continue 'normally' or even that culture might be 'rebuilt'—as if the rebuilding of culture were not already its negation—is idiotic. Millions of Jews have been murdered, and this is supposed to be seen as an interlude and not the catastrophe itself. What more is this culture waiting for?"[28] As little as one can transfer this American note of 1944 directly to the German situation in 1950–1951, when Adorno actually and practically stood before the task of reorienting himself, it does nevertheless provide a foothold in two respects. For one thing, it makes it clear how much Adorno was preoccupied with questions of cultural continuity. Yet on the other hand, it shows unmistakably what it was about the state of contemporary experience that served as the actual and insurmountable obstacle to doing as much as imagining a cultural continuity. This was the annihilation of the Jews under National Socialism, and thus also the death threat aimed at Adorno himself in the context of the most recent German history—or, as the code word for this will soon read—"Auschwitz."

The view propounded in this chapter, in the context of the study of contemporary history, is that in his twenty years of functioning in the Federal Republic, Adorno not only became a cultural bellwether, but that he also—and decidedly—stood for a political project: for the critical confrontation with the National Socialist past, which he explained as a central precondition for the build-up and stabilization of a democratic culture. How was this hardly predictable result possible, when one recalls Adorno's highly esoteric starting point at the beginning of the 1950s? What intellectual methods and practices that make this development explicable were put into effect? And, finally, within which time frame did Adorno become a leading political figure in the Federal Republic, a society whose fabric was still to a very high degree threaded with pre-democratic traditions? It is the thesis of this chapter that the concentric answer to all of these questions becomes possible, if one concentrates on that aspect of Adorno's postwar production that I here call "political culturalism."[29]

This term does not indicate so much the pursuit of political goals by cultural means in general, but rather it is directed toward a particular configuration. Intended is the curious and by no means easily understandable ambiguity present in the mutual mir-roring of culture and politics, as it was installed in the German political–moral situation after National Socialism, and to which Adorno, for his part, knew how to react in a quite remarkable way. It would be mistaken, as we have already seen, to assume that a direct cultural–political strategy was given from the beginning. What must rather be supposed is a more complex process, which—though it had been earlier prepared in Adorno's manner of thinking and writing—first took clear form during the period of re-emigration. In its center stood the traumatic experience of the annihilation of the Jews, which had become ever more immediate since the 1940s for even the far-removed emigrants, and which in the 1950s took on the form of a deeply painful memory. What can be generalized about the generation of survivors and emigrants followed quite a distinctive course for Adorno. For him, the traumatic memory seems to have continued to lead a sort of underground existence, that in a repeated yet sporadic fashion—eruptively and at first nonpolitically—made its appearance, until it finally crystallized into a political argument, which as if bursting through the surface, entered the political public sphere of the Federal Republic.

This is of course only a hypothetical formulation, which ascribes an unrealistic finality to the relationship between culture and politics, but there is in fact a point in Adorno's postwar intellectual development that represent, both in terms of logic and time, an exceptional telling event. The reference is to the well-known speech of 1959, *The Meaning of "Working through the Past."*[30] If there is anywhere tangibly present in Adorno's work a direct political message, suitable for a political manifesto, then it is here in this rhetorical miniature, as one could call it. What is noteworthy about this text, which was published shortly after the speech was given, is how such a great effect could spring from such a limited format, and from a mere occasional piece at that. No less noteworthy—and significant for our objective in this inquiry—is, conversely, that this unquestionably political text nonetheless was and remained a "genuine Adorno," with everything that pertains to it. One finds free and improvised speech, as well as the point of departure in subjective experience and its political dramatiza-tion. Present also is the popularization of subtle methodological and social–scientific problems. One might go so far as to claim that a dramatic or even musical form underlies the thematic outline, something along these lines: (1) the "cold forgetting" of the victims is lamented and described as a form of the afterlife of National Socialism (2) an association to the lack of a democratic culture is set forth (3) the pathology of forgetting is subjected to social-psychological analysis (4) the speech rises to a threatening picture of the total context of deceit, and (5) the excitement ebbs with a delineation of the task of "democratic pedagogy."

My intention here is not to analyze Adorno's 1959 speech; I aim rather to fix it as an unambiguous place marker from which to look back on the 1950s. If one recalls the location of this lecture—Adorno spoke before the Coordinating Council for Jewish–Christian Cooperation—and if one further recalls the moment in time—just a few weeks before the infamous and consequential scrawling of swastikas during the Christmas holidays, an incident that began in Cologne, but was followed by an entire series—one could almost say that Adorno had a prophetic gift. Here, however, we must take his rejection of the concept of genius at its word and rely instead on the objective context. In the historical reconstruction of the political culture of the

Federal Republic, the turn from the 1950s to the 1960s is nowadays seen as a decisive caesura, at least when the "politics of the past" is the focus. While the German relationship to the Nazi past prior to this caesura was characterized—in a relatively abrupt break with the immediate postwar years—by a wall of intentional silence, but also by the quiet but highly effective reintegration of National Socialist perpetrators as well as collaborators, this wall begins to crumble in the 1960s. Legal criminal prosecution is once again intensified, and the public sphere is declared a forum with jurisdiction over political appeals concerning "reworking the past." What the philosopher Hermann Lübbe more excused than explained with the term "communicative silence," the political scientist Helmut König characterized with the more accurate concept of an extenuating "double strategy."[31]

If Adorno's 1959 speech was the historical prelude to the transition to an active and enlivening phase of a decidedly political "coming to terms with the past"—how do the 1950s look when viewed from this perspective? It is my thesis that Adorno himself also employs a "double strategy," though not one, to be sure, that was oriented toward denial and silence, but rather toward a subtle form of Enlightenment, that in typical fashion proceeded indirectly. His discourse was "culturalistically" grounded and therefore spoke directly neither of politics nor to it, but used weaker and more esoteric themes, media and methods. His texts appeared as feuilleton publications and in periodicals; they were presented as a literary interpretation or as an essay that attempted to unlock difficult questions of music theory. In short, after his return, Adorno quickly developed into a no less multifaceted than inventive impresario of a new way of writing, even of a regular rhetoric, which I have here called "culturalistic politics" or "political culturalism." Its by no means obvious goal was the build-up and stabilization of democratic culture in the postwar society; its moral core was a demand for truthfulness in dealing with the most recent past; but its means were wilful, in part oversubtle, in part enigmatic, and yet always comprehensive and penetrating cultural interpretations.

In conclusion, I would simply like to point to rather than to analyse some telling examples out of Adorno's almost overwhelming publicistic activity of the early 1950s, whereby the sheer diversity of this prolific activity in itself already provides one explanation for the attention that Adorno quickly drew to himself. Evidently contributing to the same effect was the remarkable multivocality of this activity, clearly tied to Adorno's Enlightenment strategy. If we add in the obvious concentration of this indirect politics on the esoteric level of bourgeois high culture, then we will have assembled some of the factors that led to Adorno's success. All this together, and not least, as well, the limits Adorno's consciously or unconsciously imposed on his criticism, his many-purposed embedding of the political message in a esoteric medium—its disguise, if you will—comprised "political culturalism." If highly eloquent ambivalences belonged to this distinctive strategy of Enlightenment, something of this sort can already be seen in the book that was just submitted at the time of Adorno's return in 1949. In the *Philosophy of New Music*, there is alongside—or, better, underneath—the confrontation between Schönberg and Stravinsky a broader and much more effective level of argumentation, upon which the actual political message of the "new music" is at once hidden and transmitted. Although the identification with the Schoenberg School is already present in Adorno, what might be called its "political charge" is loaded only in the explosive final section of the Schoenberg chapter. And even more significant here is the fact that this loading takes the form of a disguise.

The background, so rich in presuppositions, the tableau full of aesthetic and historical reflections from which this hidden political tendency can be said to hatch, cannot be detailed here. Included are, for example, the methods of immanent art criticism[32] as well as the historical movement of the "new music" to distance itself from classical harmonics and the idealist compositional aesthetic. "Today the only works that really count are those that are no longer works at all"[33] reads the radical aphorism. Included as well is the assumption of the vulnerable character of art forms in the historical–societal context, in other words something like an aesthetic of genius diverted into the social sphere: "The forms of art reflect the history of man more truthfully than do documents themselves."[34] Through the radical negation of classical form, which at the same time allows its dynamic center—musical exposition—to become total, "new music" becomes the bearer of something true from the standpoint of the philosophy of history.[35] In this connection, it is interesting that the dogmatic tendencies of the twelve-tone technique are on the one hand emphasized because they represent a threat to the freedom of artistic expression, but that, on the other hand, Adorno insists, in view of the last phase of Schoenberg's work, on the objective truth contents of this technique, which has freed itself of this danger. The concluding judgment is emphatic and amounts to nothing less than an equation in philosophy of history of "new music" with the objective world spirit, although this seems to have removed itself into social conditions. Hegelian in any case is the mode of speech that Adorno employs quite without embarrassment: "Works of art—like all precipitates of the objective spirit—are the object itself. They are the concealed social essence, quoted as the phenomenon,"[36] inasmuch as the "gestures of the works of art are objective answers to objective social configurations [. . .] yet the more precisely the work of art gives answer to the heteronomy of society, the more it becomes estranged from the world[. . .]however, it assumes a tense position against the horrors of history."[37]

Adorno's immanent interpretation of "modern music" comes into its own first through the drawing of a direct parallel between its radical negation of form and the state of affairs in the contemporary world, that is described as no less negative. It is drawn with the help of two tropes. First, it proceeds through the privileging of the art form of the fragment, which follows Walter Benjamin's distinction between the auratic and fragmentary, that is, modern, work:

> Modern music absorbs the contradiction evident in its relationship to reality into its own consciousness and form. Through such action it refines itself as a means of perception[. . .]. Its depth is that of a judgment pronounced against the negative aspects of the world. The basis for judgment in music, as a cognitive force, is aesthetic form[. . .]. Modern art permits the contradiction to remain, revealing the original foundation of its categories of judgment—that is, of form[. . .]. It is only in a fragmentary work that has renounced itself that the critical substance is liberated.[38]

Second—and this for us the decisive idea because it shows the political mission and its encoding both at work—it proceeds through the metaphor of the message in a bottle:

> The shocks of incomprehension, emitted by artistic technique in the age of its meaninglessness, undergo a sudden change. They illuminate the meaningless world.

Modern music sacrifices itself to this effort. It has taken upon itself all the darkness and the guilt of the world. Its fortune lies in the perception of misfortune; all of its beauty is in denying itself the illusion of beauty. It dies away unheard, without even an echo. If time crystallizes around that music which has been heard, revealing its radiant quintessence, music which has not been heard falls into empty time like an impotent bullet. Modern music spontaneously aims toward this last experience, evidenced hourly in mechanical music. Modern music sees absolute oblivion as its goal. It is the true "message in a bottle," the surviving message of despair from the shipwrecked.[39]

If one compares this explosive conclusion of the Schoenberg chapter with the passages quoted from the essay *Cultural Criticism and Society*, one is struck by the exhortative tone and the substitutive gesture of the *Philosophy of New Music*. But above all, its "elevated speech" seems to stand in curious contrast to its lack of clarity, to the portentuous over-generality, if you will, with which the charges against everything "wicked," and against the "darkness and guilt of the world" are laid. *Minima Moralia* is much more clearly and discursively concerned with the "million-fold murder of the Jews," and yet one could still maintain that this more concrete naming of the "catastrophe" is hidden, so to speak, in the winding labyrinth of the partly intellectual–historical, partly personal *Reflections from Damaged Life*. In this context, the concise final sentence of *Cultural Criticism and Society*, which was misunderstood as a prohibition of lyric poetry and for which Adorno won early fame, offers quite a different reading. Now, enigmatically and yet momentous with meaning, the codeword "Auschwitz" comes into play. At the very end of a highly abstract discourse on cultural theory, it surfaces with shocking suddenness. It does not arise out of the argument, not to speak of an organic development. Quite to the contrary, it blows the entire discourse apart. In the process, it is precisely high and sublime culture that appears as a crust, a shell, that shatters under the accumulated pressure of a reality that is only a memory, but yet much more: an oppressive reality first of all because it appears to be delivered up to collective forgetting.

The continuing operation of this enigmatic constellation in Adorno's writings of the 1950s can only be hinted at here. It is the concluding point of *Characteristics of Walter Benjamin*, which gave impetus to Adorno's momentous reception in 1950, just as it appears at the end of *Notes on Kafka* in encoded and yet directly thematic form.[40] Here as well as in *Prisms* (1955) it could prove to be the key, not only to the breathtaking productivity of Adorno as a publicist, but also to the lively reception that his essays and cultural critical interventions received, in comparison to his academic writings in the narrower sense. To round off this study with the concentration on music that was so typical of the first period of Adorno's re-emigration, I close with some comments on the detailed essay in which Adorno, in 1953 in the journal *Neue Rundschau* once again took up the theme of "new music."

What may have been intended as a eulogy for the master of twelve-tone composition, who had died in 1951, adds noteworthy new accents to the monographic first work of 1949. Perhaps because they entered into clandestine contact with the Auschwitz code, these new accents began a subterranean restructuring of the entire historical tableau in which the development of "new music" had been embedded. Although Schoenberg still appeared as a musical revolutionary, Adorno now almost seemed to attach more importance to the gigantic role that permitted, with the destruction of traditional harmony, the archaic and conservative elements to come to the fore. This turned Schönberg into a "conservative revolutionary": "It combines aesthetic avant-gardism with a conservative mentality. While inflicting the most deadly blows on authority through his work, he seeks to defend the work as though before a hidden authority and ultimately to make it itself the authority."[41] If this "new music" is the actual heir of classical musical language, as it attained perfection and at the same time began to crumble in the late works of Beethoven—"Schoenberg's spontaneous productive power executed an objective historical verdict—he liberated the latent structure while disposing of the manifest one,"[42]—then an analogous heightening can also be discerned in Schoenberg's creative biography and translated into the social, into the catastrophe of the time. "His music expands like a giant, as though the totality, the 'great storm,' were about to emerge from self-oblivious subjectivity, 'all alone.' "[43]

Seen from this perspective, it is only logical—and consistent in any case with regard to the cryptic grammar that I have here employed for Adorno's political culturalism—that this homage to Schoenberg—and with it the essay as a whole—ends with the hardly cryptic late work, which comes closest to the Auschwitz code, and which can even be understood as its appropriate setting to music: *Survivor from Warsaw*. And perhaps one can read the emphatic sentences with which Adorno characterizes this work as in fact a—hidden political—challenge, to have it be taken to heart, above all, where it properly belongs: the land of the perpetrators—to raise a memorial to the victims.

Oratorio and Biblical opera are outweighed by the tale of the *Survivor from Warsaw*, which lasts only a few minutes; in this piece, Schoenberg, acting on his own, suspends the aesthetic sphere through the recollection of experiences which are inaccessible to art. Anxiety, Schoenberg's expressive core, identifies itself with the terror of men in the agonies of death, under total domination [. . .]. That which the feebleness and impotence of the individual soul seemed to express testifies to what has been inflicted on mankind in those who represent the whole as its victims. Horror has never been sounded as truthfully in music, and by articulating it, music regains its redeeming power through negation. The Jewish song with which the *Survivor from Warsaw* concludes is music as the protest of mankind against myth.[44]

Here I need only add that when this work, Schoenberg's Opus 46, was first performed in Germany in the fall of 1951 as part of the Kranichstein program for New music, it was due not least to the instigation of Adorno, advancing political culturism to a new, nonverbal dimension of *Bildung*.

Notes

1. In Germany, in August of 2003 alone, the following works appeared: Stefan Müller-Doohm, *Adorno. A Biography* (Frankfurt: Suhrkamp, 2003); Detlev Claussen, *Theodor W. Adorno. A Last Genius* (Frankfurt: S. Fischer, 2003); Lorenz Jäger, *Adorno. A Political Biography* (Munich: Deutsche Verlags-Anstalt, 2003).
2. *Theodor W. Adorno: Letters to his Parents 1939–1951*, ed. Christoph Gödde and H. Lenitz. (Frankfurt: Suhrkamp, 2003); *Theodor W. Adorno, Thomas Mann. Correspondence 1943–1955*, ed. Christoph Gödde and Thomas Sprecher (Frankfurt: Suhrkamp, 2002).

3. Quotations are translated from *Collected Writings 20.2* (Frankfurt: Suhrkamp, 1986), 453–466.
4. Ibid., 453.
5. Ibid., 454.
6. Ibid., 455.
7. Ibid., 455–456.
8. Ibid., 454.
9. Ibid., 458.
10. Ibid., 460.
11. All citiations ibid., 457.
12. Ibid., 459.
13. Ibid., 463.
14. Cf. Theodor W. Adorno, "Negative Dialectic," *Collected Writings*, vol. 6,4 (Frankfurt: Suhrkamp, 1990), 355.
15. Citations taken from *The Adorno Reader*, ed. Brian O'Connor (Oxford: Blackwell Publishers Ltd., 2000), 210.
16. Ibid., 199.
17. Ibid., 200.
18. Ibid., 200.
19. Ibid., 201.
20. Ibid., 202.
21. Ibid., 204.
22. Ibid., 205.
23. Ibid., 207.
24. Ibid., 207.
25. Ibid., 208.
26. Ibid., 210.
27. Cf. the extensive and commendable evaluation of the reception history by Alex Demirovic, *The Non-conformist Intellectual* (Frankfurt: Suhrkamp, 1999), beginning on p. 537.
28. Citations taken from Th. W. Adorno, *Minima Moralia: Reflections from Damaged Life*, trans. E. F. N. Jephcott (London: Verso Editions, 1978), 55.
29. Here I follow the lead of the following study, but modify its perspective: *The Intellectual Foundation of the Federal Republic: A History of the Influence of the Frankfurt School*, ed. Clemens Albrecht, Günther C. Behrmann, Michael Bock, Harald Homann, Friedrich H. Tenbruck (Frankfurt, New York: Campus, 1999). See also my essay: "Adorno and the Political Culture of the Young Federal Republic," in *Mittelweg* 36, 2 (2002), particularly the section beginning with p. 40.
30. "The Meaning of 'Working through the Past,' " reprinted in *Collected Writings*, vol. 10.2. (Frankfurt: Suhrkamp, 1986) beginning on p. 555.
31. Helmut König, *The Future of the Past: National Socialism in the Political Consciousness of the Federal Republic* (Frankfurt: Fischer, 2003), in particular 24–30. Norbert Frei offers a comprehensive historical account in his *Politics of the Past: The Beginnings of the Federal Republic and the National Socialist Past*, second edn. (Munich: Beck 1997).
32. Citations taken from Theodor W. Adorno, *Philosophy of New Music*, trans. Anne G. Mitchell and Wesley V. Blomster (New York: Seabury Press, 1973).
33. Ibid., 30.
34. Ibid., 43.
35. Ibid., 51–67.
36. Ibid., 131.
37. Ibid., 132.
38. Ibid., 124–125.
39. Ibid., 133, translation modified.

40. "Characteristics of Walter Benjamin," in: *Collected Writings*, vol. 10.1 (Frankfurt: Suhrkamp, 1986) in particular 252; *Notes on Kafka*, ibid., in particular 286–287.
41. Citations taken from *Arnold Schoenberg, 1874–1951*, trans. Samuel and Sherry Weber, in *The Adorno Reader*, ed. Brian O'Connor (Oxford: Blackwell Publishers, 2000), 284.
42. Ibid., 288.
43. Ibid., 295.
44. Ibid., 302–303.

INDEX